STRUCTURE AND BONDING

Volume 23

Editors: J. D. Dunitz, Zürich
P. Hemmerich, Konstanz · R. H. Holm, Cambridge
J. A. Ibers, Evanston · C. K. Jørgensen, Genève
J. B. Neilands, Berkeley · D. Reinen, Marburg
R. J. P. Williams, Oxford

With 50 Figures

Springer-Verlag
Berlin Heidelberg GmbH 1975

ISBN 978-3-662-15524-0   ISBN 978-3-540-37566-1 (eBook)
DOI 10.1007/978-3-540-37566-1

Library of Congress Catalog Card Number 67-11280

© by Springer-Verlag Berlin Heidelberg 1975
Originally published by Springer-Verlag Berlin Heidelberg New York in 1975
Softcover reprint of the hardcover 1st edition 1975

Contents

STRUCTURE AND BONDING is issued at irregular intervals, according to the material received. With the acceptance for publication of a manuscript, copyright of all countries is vested exclusively in the publisher. Only papers not previously published elsewhere should be submitted. Likewise, the author guarantees against subsequent publication elsewhere. The text should be as clear and concise as possible, the manuscript written on one side of the paper only. Illustrations should be limited to those actually necessary.

Manuscripts will be accepted by the editors:

Professor Dr. *Jack D. Dunitz*

Laboratorium für Organische Chemie der Eidgenössischen Hochschule
CH-8006 Zürich, Universitätsstraße 6/8

Professor
Dr. *Peter Hemmerich*

Universität Konstanz, Fachbereich Biologie
D-7750 Konstanz, Postfach 733

Professor *Richard H. Holm*

Department of Chemistry, Massachusetts Institute of Technology
Cambridge, Massachusetts 02139/U.S.A.

Professor *James A. Ibers*

Department of Chemistry, Northwestern University
Evanston, Illinois 60201/U.S.A.

Professor
Dr. *C. Klixbüll Jørgensen*

51, Route de Frontenex,
CH-1207 Genève

Professor *Joe B. Neilands*

University of California, Biochemistry Department
Berkeley, California 94720/U.S.A.

Professor Dr. *Dirk Reinen*

Fachbereich Chemie der Universität Marburg
D-3550 Marburg, Gutenbergstraße 18

Professor
Robert Joseph P. Williams

Wadham College, Inorganic Chemistry Laboratory
Oxford OX1 3QR/Great Britain

SPRINGER-VERLAG

D-6900 Heidelberg 1
P. O. Box 105280
Telephone (06221) 487·1
Telex 04-61723

D-1000 Berlin 33
Heidelberger Platz 3
Telephone (030) 822001
Telex 01-83319

SPRINGER-VERLAG
NEW YORK INC.

175, Fifth Avenue
New York, N. Y. 10010
Telephone 673-2660

Copper Proteins
Systems Containing the "Blue" Copper Center

James A. Fee

Biophysics Research Division,* Institute of Science and Technology, University of Michigan, Ann Arbor, Michigan 48105, U.S.A.

and

The Department of Chemistry, Rensselaer Polytechnic Institute, Troy, New York 12181, U.S.A.

Table of Contents

---

* Present Address.

J. A. Fee

I. Introduction

This review is concerned with the chemical and physical properties of proteins and enzymes containing three distinct and unique forms of Cu; The "blue" center or, in the nomenclature proposed by *Vänngård*, Type 1 Cu^{2+}; the "colorless" or Type 2 Cu^{2+}, common to all multi-copper oxidases which reduce molecular oxygen to two molecules of water; and the Cu associated with the 330 nm absorption band, again common to the oxidases. The purposes of the review are to assemble chemical and physical data related to the indicated types of Cu binding sites, to offer some interpretations (and occasionally re-interpretations) of experimental results concerned with structure-function relationships, and to generalize some of the information available as it concerns the structures of these unique Cu-co-ordination complexes. Special emphasis will be placed on the kinetic and mechanistic work which has been carried out on the multi-copper oxidases while the physiological roles of the various protein systems will not be of particular importance.

General Properties of the Three Types Copper

Type 1 Cu^{2+}. This Cu^{2+} complex is characterized by two unique and apparently inseparable properties: an intense multi-banded absorption envelope in the region of 600 nm and an EPR[1]) spectrum having an unusually small hyperfine coupling constant, A_{11}. The extinction coefficient of the band near 600 nm is approximately 100 times larger than that, apparent in the same spectral region, of most Cu^{2+} complexes. Indeed, these two properties signal a structural arrangement of the Cu^{2+}-protein combination which is entirely unique among Cu^{2+} coordination complexes.

The Type 1 or blue cupric center is present in all the proteins which will be discussed in this review, either alone or in combination with two other types of copper complexes.

Type 2 Cu^{2+}. This form of Cu is present in all the blue multi-copper oxidases. It is characterized by lacking sufficient optical absorption to be observed above that of the other Cu-chromophores in these molecules. Consequently it is sometimes referred to as the colorless Cu. Further, its EPR spectrum is similar to those exhibited by most small Cu^{2+} complexes. However, its presence is essential to the functioning of the multi-copper oxidases, and it has very unique chemical properties which distinguish it from Cu^{2+} bound to the non-blue Cu proteins. The Type 2 designation should therefore be reserved for classification of the types of Cu^{2+} sites observed in the blue multi-copper oxidases, and it should not be used to classify the binding sites of non-blue copper proteins which have distinctly different chemical behavior. Thus, for example, any purported analogies between

---

[1]) *Abbreviations used.* CD, circular dichroism; EPR, electron paramagnetic resonance; ENDOR, electron nuclear double resonance; PCMB, PMB, p-chloromercuribenzoate; EDTA ethylenediaminetetraacetic acid; G·HCl, guanidine hydrochloride; PPD, p-phenylenediamine; DPD, N,N-dimethyl-p-phenylenediamine; SDS, sodium dodecylsulfate.

2

Type 2 Cu^{2+} and Cu^{2+} in superoxide dismutase, benzylamine oxidase, and galactose oxidase are fallacious.

Type 3 Cu. This form of Cu is also found in all multi-copper oxidases and is essential to the reduction of oxygen. It is characterized by its ability to act as a two-electron acceptor/donor system, an absorption band at 330 nm, the lack of an EPR spectrum, and it is non-paramagnetic over a wide range of temperatures. This center appears to consist of two Cu^{2+} ions, in close proximity, which are strongly antiferromagnetically coupled.

II. Copper Proteins Containing Single or Independent "Blue" Centers

Certain species of bacteria, non-photosynthetic plant material, and chloroplasts have been found to contain relatively low molecular weight proteins having either a single or two identical and independent Cu centers. The names and known chemical and physical properties of these proteins are assembled in Table 1.

A. Azurins

The first report of a blue protein from bacterial sources was that of *Verhoeven* and *Takeda* (*1*). These workers isolated a blue pigment from *Pseudomonas aeruginosa* having a broad absorption band at 630 nm which disappeared on addition of dithionite or upon heating. This observation was followed by work of *Horio* (*2, 3*) who developed a procedure for purification and established that the pigment was a protein containing a single copper per molecule. It was tentatively named *Pseudomonas* blue protein, a name which is still used to some degree. Later, *Sutherland* and *Wilkinson* (*4*) isolated a similar protein from *Bordetella pertussis*, and suggested that all such proteins be called azurins, a name which is now widely used.

Proteins of this class which have received the most attention were isolated from four bacterial species: *Pseudomonas aeruginosa*, *Ps. fluorescens*, *Ps. denitrificans*, and *Bordetella pertussis*, although *Sutherland* (*5*) has isolated azurin from several other strains of *Pseudomonas*, *Bordetella*, and *Alcaligenes*. The near identity of azurins isolated from these different sources, has been generally assumed, and with one possible exception, that from *Ps. denitrificans* which does not bind to carboxy-methyl cellulose resin at the same pH as the other proteins (*5*), this seems to be the case. *Ambler* and *Brown* (*6*), who have elucidated the amino acid sequence of *Ps. fluorescens* azurin state that *B. bronchiseptica*, *A. denitrificans* and *faecalis*, *Ps. denitrificans* and *fluorescens* yield azurins having homologous amino acid sequences.

1. General Chemical Properties. The information assembled in Table 1 is consistent with azurins from the several sources indicated containing a single Cu atom in a protein of molecular weight approximately 16,000 and very little if any associated carbohydrate.

The optical absorption and circular dichroism spectra of *Ps. fluorescens* azurin are shown in Fig. 1. The broad, intense band near 600 nm demonstrating significant multiplicity is typical of the blue Cu-center. The ultraviolet spectrum of

Table 1. Properties of proteins containing only single or independent "blue" copper centers[a]

Source	Pseudomonas aeruginosa		Bordetella pertussis	Pseudomonas denitrificans	Pseudomonas fluorescens
Chemical Properties					
Commonly used name	Pseudomonas blue protein (3) Azurin (4)		Azurin (4)	Blue protein (211) (Azurin)	Azurin
Number of Cu	1		1	1	1
Hydrodynamic parameters	$S_{20} = 1.91\,s$ $D_{20} = 1.06 \times 10^{-6}$ (209) cm² sec⁻¹ $\bar{v} = 0.749$		$S_{20,w} = 1.59\,s$ $D_{20,w} = 1.06 \times 10^{-6}$ (4) cm²/sec	$S_{20,w} = 1.90\,s$ $D = 1.11 \times 10^{-6}$ (211)	
Molecular weight	30,000 (Cu) (3) 14,600 (aa) (210) 17,400 (S,D) (209) 16,600 (Cu) (209)		14,600 (S,D) (4) 14,700 (Cu) (4)	16,300 (S,D) (211) 16,600 (Cu) (211) 16,000 (X-ray) (212)	16,000 (Cu) (213) 13,944 (aa) (6)
Isoelectric point	5.40 (209)				
Spectral criterion of purity			$A_{625}/A_{280} = 0.47$ (4)		$A_{625}/A_{280} = 0.59$ (213)
Amino acid composition	+ (210)			Sequence (6)	Sequence (6)
Half-cysteine composition	2 (210)				3 (6)
Sulfhydryl groups[b]					1 (6)
Carbohydrate composition					<0.02% (213, 6)
Oxidation reduction potential	300 mV (pH 7) (3) 328 mV (pH 6.4) (209)		395 mV (pH?) (4)	230 mV (pH 6.8) (211)	
Spectroscopic and magnetic properties					
EPR parameters	(11)	(202)	(88)	(202, 11)	(11)
g_{11}	2.260	2.26	2.273	2.26	2.261
$g_\perp$	2.052	2.055	2.049	2.055	2.052
A_{11} (cm⁻¹)	0.006	0.006	0.006	0.006	0.0058
$A_\perp$ (cm⁻¹)	~0				~0
Optical absorption spectrum nanometers or wavenumbers (mM⁻¹cm⁻¹)	(11) 21,400 (0.27) 16,000 (3.5) 12,200 (0.39)			(211) 470 (1.3) 620 (10.5?) 770 (2.6)	(11) 21,800 (0.285) 16,000 (3.5) 12,800 (0.32)
Circular dichroism spectrum	(9) 26,800 (+) 24,500 (+) 21,400 (−) 19,200 (+) 16,100 (+) 12,400 (−)				

[a] Measured values only. For example, calculated values of $\bar{v}$ were not included.
[b] Observed only in the apo-protein.
[c] All bands probably not resolved.

Table 1. (continued)

Rhus vernicifera Latex	Horse radish root	Etiolated Mung bean (25)	Rice bran (219)	Chloroplasts Spinach	French bean (34)
Stellacyanin (24)	Umecyanin			Plastocyanin	
1	1	1		2	1
$S_{20,w} = 2.0$ s	$S_{20,w} = 1.9$ s	$S_{20} = 2.8$ s	$S_{20,w}$	$S_{20,w} = 1.72$ s	$S_{20,w} = 1.36$ s
$D_{20,w} = 0.6 \times 10^{-6}$	(26)		$= 1.98$ s	$D_{20,w} = 0.66 \times 10^{-6}$	
(22) cm^2 sec^{-1}				(33) cm^2 sec^{-1}	
$\bar{v} = 0.68$				$\bar{v} = 0.70$	
25,000 (S,D)	15,000 (S,D)	22–26,000	23,000	21,000 (S.D)	10,790 (Approach
(22)	(26)	(S,D)		(33)	sed. equil.)
20,000 (Cu)	13,800 (Cu)	22,700		10,800 (Cu)	11,000 (Gel.
(22)	(216)	(Cu)		(33)	chrom.)
9.86 (214)	5.85 (26)			< 4 (33)	
$A_{608}/A_{280} = 0.17$	A_{610}/A_{280}	A_{598}/A_{280}		A_{597}/A_{278}	A_{597}/A_{278}
(22)	$= 0.27$ (216)	$= 0.17$		$= 1.25$ (33)	$= 0.91$
+ (24)	+ (26)			+ (33)	+
5–6 (24)	3 (26)			2 (33)	1
1 (29)				2 (37)	
20% (24)	3.7% (26)	30%	Present	5.4% (33)	
184 (7.1) (64)	283 (pH 7) (217)			Broken chloroplasts 370 (pH 7) (33)' Intact chloroplasts 343 (pH 7.8) (38) Chlorella – 390 (pH 7) (31)	
(27)	(24)	(23)			(36)
2.287	2.30	2.317			2.226 Chenopodum
2.077, 2.025	2,06, 2.03	2.05			2.053 Album
0.0035	0.0033	0.0035			0.0063
0.0029, 0.0057;	0.0047, < 0.001				< 0.0017
(22), (27)					(33)
450		450	450		490 Spinach
608 (4.03)		598	600		597 (4.9)	597 (4.5)
850		800–850			780
(215)	(218)				(36)
33,000 (+)	24,600 (−)				22,100 (−) Chenopodum
22,400 (−)	22,150 (−)				16,700 (−) Album
19,000 (+)	19,100 (+)				12,700 (+) ORD)^c)
16,500 (+)	16,500 (+)				
12,800 (−)	12,700 (−)				
11,000 (+)					

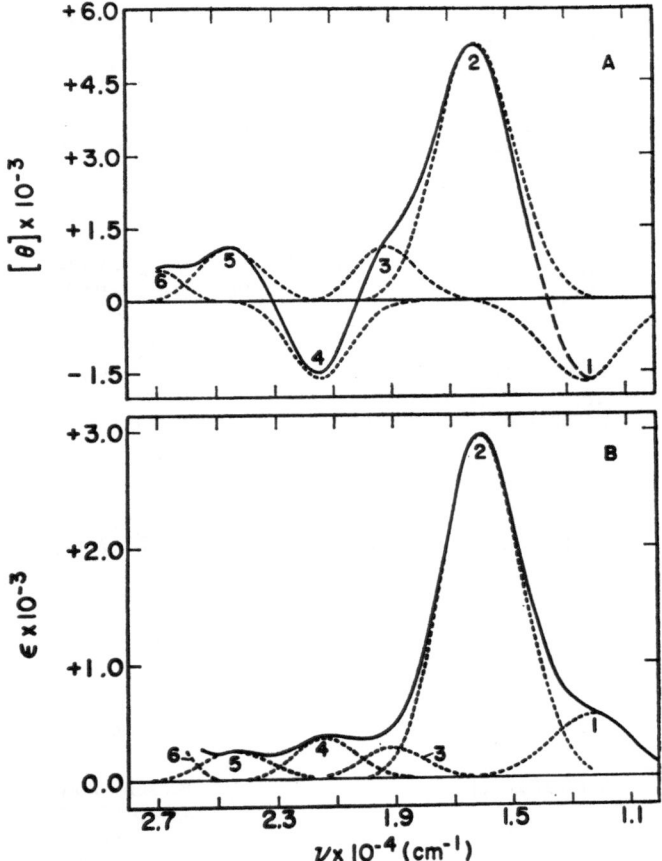

Fig. 1. Visible circular dichroism (A) and absorption (B) spectra of Pseudomonas aeruginosa azurin. The solid line corresponds to the observed spectral envelope, and the dashed lines are individual Gaussian bands the sum of which represent the experimental spectra. The CD bands are listed in Table 1. [Taken from *Tang, Coleman,* and *Myer*; Ref. (9)]

azurin is unique in having a sharp maximum at 292 nm and fine structure in the region of 280 nm (7). The available spectroscopic and EPR parameters of the Cu^{2+} of azurins are assembled in Table 1.

The amino acid sequence of *Ps. fluorescens* azurin, Fig. 2, determined by *Ambler* and *Brown* (6) reveals information which is fundamental to any understanding of the physical properties of the blue center. During the course of this work a number of observations suggested that azurin possessed a very compact structure and that the metal played some role in maintaining this structure (6). For example, the native protein was found to be completely resistant to proteolysis by trypsin, chymotrypsin, or subtilysin B while, in contrast, the apoprotein was readily digested by these enzymes. Another manifestation of the compactness of the protein structure was the failure to detect significant amounts of tryptophan (6) in undigested material by the chemical method of *Spies* and *Chambers* (8)

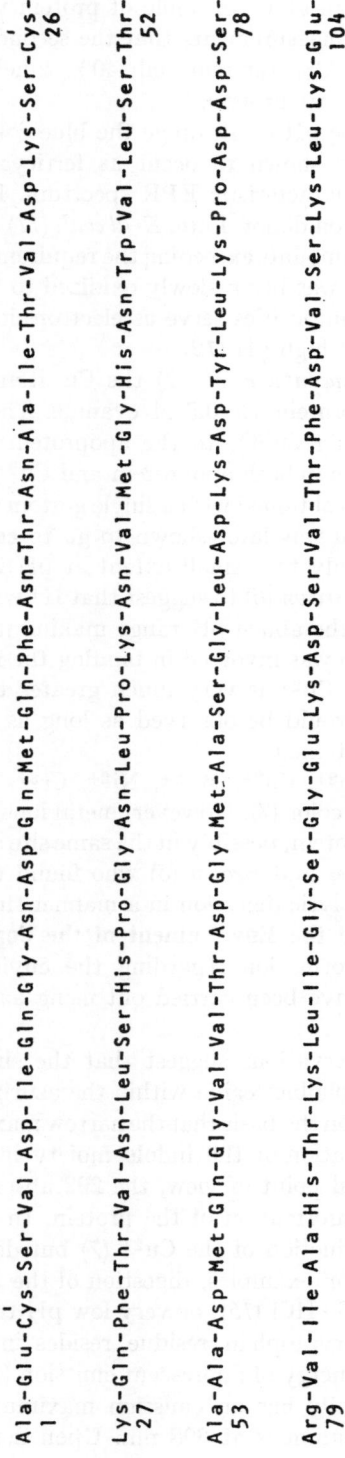

Ala-Glu-Ser-Val-Asp-Ile-Gln-Gly-Asn-Asp-Gln-Met-Gln-Phe-Asn-Thr-Asn-Ala-Ile-Thr-Val-Asp-Lys-Ser-Cys-
1   26

Lys-Gln-Phe-Thr-Val-Asn-Leu-Ser-His-Pro-Gly-Asn-Leu-Pro-Lys-Asn-Val-Met-Gly-His-Asn-Trp-Val-Leu-Ser-Thr-
27   52

Ala-Ala-Asp-Met-Gln-Gly-Val-Val-Thr-Asp-Gly-Met-Ala-Ser-Gly-Leu-Asp-Lys-Asp-Tyr-Leu-Lys-Pro-Asp-Asp-Ser-
53   78

Arg-Val-Ile-Ala-His-Thr-Lys-Leu-Ile-Gly-Ser-Gly-Glu-Lys-Asp-Ser-Val-Thr-Phe-Asp-Val-Ser-Lys-Leu-Lys-Glu-
79   104

Gly-Glu-Gln-Tyr-Met-Phe-Phe-Cys-Thr-Phe-Pro-Gly-His-Ser-Ala-Leu-Met-Lys-Gly-Thr-Leu-Thr-Leu-Lys
105   128

Fig. 2. The amino acid sequence of azurin from *Pseudomonas fluorescens*. [From *Ambler* and *Brown*; Ref. (6)]

whereas in digested material one residue per mole of protein was found. *Tang et al.* (*9*) have estimated from CD measurements that the secondary structure of azurin consists of approximately 23% random coil, 40% α-helix, and 37% β-structure. This is unchanged in the apoprotein.

Maria (*10*) observed that in the pH 8—12 range the blue color of azurin was slowly bleached. A reduction was shown to occur as ferricyanide completely restored the blue color and the characteristic EPR spectrum. It was suggested that SH groups served as the electron donor. Later *Brill et al.* (*11*) observed that reducing equivalents were present in amounts exceeding the requirement to reduce the Cu^{2+}. It was suggested that water was being slowly oxidized to H_2O_2 or O_2, but it is more likely that endogenous impurities serve as electron donors in this and other reductions of Type 1 Cu^{2+} at high pH (*12*).

As first demonstrated by *Yamanaka et al.* (*7*) the Cu atom can readily be removed by dialysis of reduced protein vs. 0.5 M cyanide. The native protein could be restored by simply adding $CuSO_4$ to the apoprotein; the velocity of reformation was apparent first order in both apoprotein and Cu^{2+} concentrations, and full blue color development was obtained with a single g-atom Cu^{2+} per mole of apoprotein (*7*). This recombination was later shown to go to completion in the pH range 3.1—9.0 and to occur only to a small extent at pH 2.6 or 10 (*6*), an observation which led *Ambler* and *Brown* (*6*) to suggest that H^+ was not competing with Cu^{2+} for the binding site in the above pH range making it unlikely that a group with a pK_a between 4 and 8 was involved in binding the metal. However, if the overall binding constant of Cu^{2+} is very much greater than that of the proton then no effect of acidity would be observed as long as the proper conformation of the protein was maintained.

Other metal ions including Co^{2+}, Co^{3+}, Mn^{2+}, Ni^{2+}, Cr^{3+}, Fe^{2+}, and Fe^{3+} were shown not to restore the blue color (*7*). However, metal ions other than Cu^{2+} appear to bind firmly to the apoprotein, possibly in the same site as Cu^{2+}. The evidence for this was given by *Ambler* and *Brown* (*6*) who found that Co^{2+} would protect the apoprotein from proteolytic digestion in a manner similar to Cu^{2+}.

2. Individual Amino Acids and the Environment of the Copper. Several experiments designed to obtain information regarding the environment of the Cu^{2+} and its immediate ligands have been carried out using azurin *Ps. fluorescens*.

Tryptophan. A number of observations suggest that the single tryptophan residue occupies a strongly hydrophobic region within the azurin molecule. This was first suggested by *Teale* [cf. (*6*)] on the basis that the narrow maximum at 292 nm arises from a well known perturbation of the indole moiety in a hydrophobic environment (*13*). From a practical point of view, the 292 absorption band can serve as a probe of the native conformation of the protein. In this regard, the band remains upon removal or reduction of the Cu^{2+} (*7*) but denaturing conditions lead to its disappearance; for examples, digestion of the apoprotein with proteases (*14*), dissolution in 6 M G · HCl (*15*), or very low pH treatment (*15*). A further property suggesting the tryptophan residue resides in a hydrophobic environment is its unusually high energy of fluorescent emission (*15*). Tryptophan in hydrophilic surroundings generally has an emission maximum near 350 nm while in azurin the emission maximum is at 308 nm. Upon denaturation with acid or 6 M G · HCl this shifts to 350 nm. Dissolution of the acid denatured protein in 2% SDS restores the emission maximum to ~310 nm (15), presumably because the indole moiety is surrounded by the alkyl chains of the SDS molecules.

The fluorescence emission intensity from holoprotein is approximately 40% that observed with apoprotein (15). Addition of Cu^{2+} or Hg^{2+} to apoprotein caused a stoichiometric decrease in fluorescence. PMB, however, had no affect on the emission intensity (15). Cu^{2+}, Hg^{2+}, and Ag^+ also have some quenching effect on the luminiscence of the tryptophan. While the single tryptophan residue serves as a "reporter" of the native conformation of the protein the following considerations suggest that it is probably not a direct ligand to the Cu^{2+}: (a) Removal of the Cu^{2+} does not change the UV absorption characteristics of the system which are largely dominated by this residue. (b) A complete quenching of both fluorescence and luminescence emission would be expected (16). (c) The exceptionally high pK_a of the indole N-H (17) would not exclude the indole anion as a ligand but does suggest that its coordination would be extraordinary.

Tyrosine. Azurin contains two tyrosine residues (6). *Finazzi-Agro et al.* (15) and *Avigliano et al.* (18) have titrated azurin and apoazurin to very high pH and measured changes in ΔA_{295}. Anomalous behaviour such as hysteresis effects and unusually large changes in absorbance at 295 nm were observed. Estimated pK_a values are ~12.3 and ~12.5 for the tyrosines of azurin and apoazurin respectively. However, Cu^{2+} does not appear to be bound to tyrosine as there are no differences between the UV spectral properties of apo- and holoproteins.

Cysteine. Some chemical observations suggest but do not definitely prove that the single sulfhydryl group of azurin may be a direct ligand to Cu^{2+}. This was suggested by the inhibition of the reconstitution process by the sulfhydryl binding agents Hg^{2+}, Ag^+, and PMB (15, 19), all of which bind to apoprotein in 1:1 stoichiometry. It was argued that the inhibition is effected by a competitive binding of -S⁻ to Cu^{2+} and to metals having a high affinity for sulfhydryl. Attempts to block the cysteine by alkylation have been unsuccessful apparently because the single cysteine is flanked by very bulky side chains; Tyr, Met, Phe, Phe on the N-terminal side and Thr, Phe on the C-terminal side, which prevent reaction of the SH group even under strongly denaturing conditions (Fig. 2) (6).

Water. *Koenig* and *Brown* (20) and *Boden et al.* (21) have studied the paramagnetic contribution of azurin bound Cu^{2+} to the relaxation rate of solvent protons and concluded that Cu^{2+} probably does not have an open coordination position to bind water. This conclusion would be consistent with the apparent inability of azurin to form stable anion-Cu^{2+} complexes.

B. Stellacyanin, Umecyanin, and Mung Bean Blue Protein

Each of these proteins is isolated from non-photosynthetic plant tissue. Stellacyanin was originally isolated by *Omura* (22) from the exudate of the lacquer tree *Rhus vernicifera*. Umecyanin is found in horse radish roots (23), and Mung bean blue protein is isolated from etiolated Mung bean plants. When properly purified these have no oxidase activity but can be reduced by various electron donors. The information in Table 1 suggests a single Cu atom per molecular weight of approximately 15—20,000 daltons. Stellacyanin (24) and Mung bean protein (25) contain a large amount (20—30%) of carbohydrate while umecyanin has only 3.7% (26). These proteins have no known or suspected biological function.

pH *Effects in Stellacyanin. Peisach et al.* (24) found that the EPR spectrum of stellacyanin was unaltered in the pH region 8.3—2.0, however, above 8.3 in the range 8.3—11.5 a transition to a "denatured" or biuret type of Cu^{2+}-complex occurred. An EPR spectrum showing superhyperfine structure arising from several N-atoms was observed, and it was concluded that the uniquely bound Cu^{2+} must have at least four N-atoms as primary ligands. Later work by *Malmström et al.* (27), however, showed that the transition between the biuret-type EPR spectrum and the native form of Cu^{2+} was not reversible; excluding any deductions regarding the nature of the immediate liganding atoms of Cu^{2+} in the native form of the protein.

The ligands of Cu^{2+} bound to stellacyanin have been probed with the electron nuclear double resonance (ENDOR) technique by *Rist et al.* (28). These authors have drawn the following conclusions about the immediate structure of the Cu^{2+} complex of stellacyanin: (a) It is not readily accessible to solvent water molecules, consistent with previous observations of *Peisach et al.* (24); (b) More than one set of N-atoms contributes to the ENDOR signal; and (c) The unpaired electron of the Cu^{2+} ion is quite strongly coupled to at least one proton in the oxidized protein which in the native oxidized form is not readily exchangeable with solvent protons.

Morpurgo et al. (29) have successfully removed and replaced the Cu^{2+} in stellacyanin. The procedures developed were used by *McMillin et al.* (30) to prepare a Co^{2+} analog of this protein. The optical spectrum of Co^{2+} stellacyanin shows weak $d \rightarrow d$ transitions in the visible region but exhibits an intense absorption near 350 nm which has been assigned as a sulfur $\rightarrow$ metal charge transfer transition. While this would seem to be good evidence that Co^{2+} is coordinated to sulfur and by analogy Cu^{2+}, it remains to be shown that Co^{2+} binds the same ligands as Cu^{2+}.

C. Plastocyanin

This protein is found exclusively in the chloroplast where it is involved in electron transfer from Photosystem II to Photosystem I. Plastocyanin has been isolated from a number of green algae: *Chlorella ellipsoidea* (31), and *Chlamydomonas rheinhardi* (32), as well as from spinach (33), French bean (34), parsley (35), and *Chenopodium album* (36).

There is some confusion in the literature concerning the molecular weight of the protein from the various sources. The early paper of *Katoh et al.* (33) on the physical properties of spinach plastocyanin indicated a molecular weight of approximately 21,000 based on sedimentation and diffusion measurements and a minimum molecular weight based on Cu analyses of 10,800 daltons. These data suggested a molecule containing two g-atoms Cu per 21,000 daltons. *Blumberg* and *Peisach* (36) report results indicating the protein from *C. album* has a molecular weight of 11,500. *Gorman* and *Levine* (32) report a molecular weight of 13,000 (gel filtration) for *C. rheinhardi* plastocyanin but do not report the percentage of Cu. *Milne* and *Wells* (34) have studied the properties of French bean plastocyanin showing that the plastocyanin has a molecular weight of 10,600 and contains one g-atom of Cu. Its amino acid composition is strikingly similar to that of spinach (*33*) when the latter is expressed in terms of residues per 10,800 daltons. The authors suggest that the spinach protein consists of a dimer of ~11,000 dalton subunits. Therefore, it would seem in general, that plastocyanin has a molecular weight near 10,500 and binds one g-atom Cu for each mole protein. Spinach plastocyanin is unique in that it appears to form a fairly strong dimeric complex. Since French bean and spinach plastocyanins have essentially identical spectroscopic properties, the dimerization does not appear to affect the properties of the Cu-binding site, and there is no interaction between the two binding sites.

Katoh and *Takamiya* (*37*) observed that stoichiometric amounts of Hg^{2+} or Ag^+ would decolorize a solution of plastocyanin. The velocity of the decolorization was greatly enhanced in the presence of denaturant. PCMB, having a general specificity for sulfhydryl groups also decolorized plastocyanin. Apoprotein could be prepared by precipitation at pH 2.0 in the presence of citrate and 50% ammonium sulfate. Apoprotein so obtained, in contrast to native protein, was found to contain one freely reacting SH group, Reconstitution could be effected by simply adding Cu^{2+} at pH 7, and this was blocked by PCMB. *Katoh* and *Takamiya* (*37*) interpreted these results as indicating that SH was involved in coordinating Cu^{2+}. Recently, *Graziani et al.* (*35*) have repeated Katoh's work with parsley plastocyanin obtaining identical results. In this case, as with azurin and stellacyanin it is not known whether mercurials decolorize by competitively removing a direct ligand to Cu^{2+} or by binding near the Cu^{2+} concomitantly changing its physical environment.

The sequences of plastocyanins from the green alga *Chlorella fusca* (*223*) and the potato plant *Solanum tuberosum* L. (*224*) have been determined. As can be noted by comparing the limited regions about the single cysteine residue present in the two plastocyanins and azurin there is not a true homology, but on the basis of the chemical similarities between the several proteins *Kelly* and *Ambler* (*223*) have suggested a functional similarity in this region of the sequence.

```
                                        SH
     Azurin (Fig. 2)  Glu—Tyr—Met—Phe—Phe—Cys—Thr—Phe—Pro
                      107                               115

     Alga    (223)    Thr—Tyr—Gly—Tyr—Phe—Cys—Glu—Pro
                      78                           85

     Potato (224)     Thr—Tyr—Thr—Phe—Tyr—Cys—Ala—Pro
                      79                           86
```

The redox potentials of a number of plastocyanins are given in Table 1, but only spinach plastocyanin has been studied at more than one pH. *Katoh et al.* (*33*) report that above pH 5.4 the redox potential is constant at 370 mV while below pH 5.4 the potential increases approximately 60 mV for each unit of pH. This is formally rationalized by the equilibrium

$$
\begin{array}{cc}
\overset{L}{\underset{|}{P - Cu^{2+}}} + \bar{e} & \underset{H^+}{\overset{}{\rightleftharpoons}} \quad \overset{LH^+}{P - Cu^+}
\end{array}
$$

where L is a direct ligand to Cu^{2+} which when reduction occurs above pH 5.4 remains bound to Cu^+ and below pH 5.4 binds a proton instead of Cu^+. However, the proton could bind to any ionization linked to the reduction of the Cu^{2+}. *Malkin et al.* (*38*) using the ferri/ferrocyanide couple report that the redox potential of spinach plastocyanin is 340 mV in the chloroplast and 370 mV in aqueous solution. This difference may be important in assigning the position of plastocyanin in the photosynthetic electron transport chain. However, the observation of *Graziani et al.* (*35*) that ferrocyanide binds very strongly to parsley plastocyanin should be considered carefully in evaluating redox potentials.

Other observations relevant to possible ligands to the Cu^{2+} are: (a) None of the plastocyanins analyzed contains tryptophan or arginine. (b) The fluorescence of tyrosine in parsley plastocyanin is partially quenched but there is no evidence that tyrosine is bound to Cu^{2+}. (c) *Spinach* and *French* bean plastocyanin contain two histidine residues while parsley plastocyanin appears to have only one. (d) *Blumberg* and *Peisach* (*36*) report a very small effect of plastocyanin bound Cu^{2+} on the magnetic relaxation properties of H_2O, suggesting H_2O does not ligand the Cu.

III. Multicopper Oxidases

In Section II we discussed the properties of certain proteins containing single or isolated and independent blue copper centers. With the exception of plastocyanin the biological function of these proteins is not known, but it is very likely that they participate in single electron transfer reactions.

The isolated Type 1 Cu^+ is insensitive toward reoxidation (*39*) by molecular oxygen. This important and general restriction on the reactivity of these ions is removed in a number of enzymes which, in addition to the blue center, contain several other Cu per molecule (*39*). The means by which Type 1 Cu^+ is reoxidized by molecular oxygen in these proteins is a fascinating and yet unsolved problem.

In the following discussion we will consider the properties of three different classes of this general type of enzyme: laccases, ceruloplasmin, and ascorbic acid oxidase. Pertinent data are assembled in Table 2. Each of these enzymes catalyses the general reactions:

$$2\ AH_2 + O_2 \longrightarrow 2\ A + 2\ H_2O \qquad 2 \times 2e^- \text{ reduction}$$

$$4\ A^- + 4\ H^+ + O_2 \longrightarrow 4\ A + 2\ H_2O \quad 4 \times 1e^- \text{ reduction}$$

where AH_2 requires two electrons to be oxidized to A.

In the case of a one electron donor (A^-), for example, ferrocyanide, which is a poor to moderately good substrate for some of these enzymes, the catalyst is clearly coupling a one electron oxidation of substrate to a four electron reduction of molecular oxygen. This is a general property of the multi-copper oxidases.

As indicated in the Introduction, there are at least three different forms of Cu in all multi-copper oxidases which have been appropriately examined. These have been classified as Type 1 Cu^{2+}, Type 2 Cu^{2+}, and Type 3 Cu; the relative proportions of these is fixed within any particular native enzyme.

A. Chemical and Physical Properties of Laccases

Laccases are enzymes which catalyze the oxidation of p-diphenols according to the reaction

$$2 \quad \begin{array}{c} OH \\ \\ OH \end{array} + O_2 \longrightarrow + 2 \begin{array}{c} O \\ \\ O \end{array} + 4 H_2O \; .$$

Actually they catalyze the oxidation of a variety of oxidizable materials (40—45), however, the Enzyme Commission has chosen to designate these enzymes as diphenol: O_2 oxidoreductases (EC 1.10.3.2).

The name laccase was first given to the enzyme by G. *Bertrand* (46) who studied its activity in crude form from the latex of the lac tree *Rhus succedanea*, and similar enzymes have been found in the latex of various other Asian lac trees (47). The particular enzyme from this source receiving the most attention has been that isolated from *Rhus vernicifera*. Another source of enzymes having properties similar to lac tree laccases are the white rot fungi. *Fåhraeus* and collaborators (48—52) have developed methods whereby a laccase can be obtained in large quantities from the mycelium of the fungus *Polyporus versicolor*. This enzyme, termed fungal laccase, is probably the best characterized of all enzymes of this type, and it will be discussed first.

1. *Polyporus versicolor* Laccase. There are two electrophoretic forms, A and B, of fungal laccase (50—52) which otherwise seem to be identical as enzymes.

The general properties of this protein are summarized in Table 2. Four copper atoms are bound per molecular weight 62—64,000, and the work of Butzow (53) suggests that if the enzyme is polymeric it does not consist of subunits which are easily dissociated. Indeed it appears that this protein, like transferrin, (MW ~80,000), may consist of only one polypeptide chain [cf. (54)].

There has been relatively little work done on the chemical properties of the individual amino acids. *Briving* and *Deinum* (55) have examined the reactivity of the cysteine SH groups. The native oxidized enzyme has no reactive SH while the enzyme denatured under anaerobic conditions in the presence of appropriate trapping agents for Cu^{2+}, exhibits a single SH group. The remaining cysteine appears to exist in two disulfide bonds.

Also assembled in Table 2 are the spectroscopic and magnetic properties of the associated Cu chromophores. The characteristic features of the optical absorption spectrum are an intense absorption band at 610 nm ($\varepsilon = 4900$ M^{-1} cm^{-1}) and a band near 330 nm in the difference spectrum of oxidized minus reduced enzyme ($\Delta\varepsilon = 2800$—3000 M^{-1} cm^{-1}) (56). The absorption band at 330 nm is unique to the multicopper oxidases (39) while the intense multi-banded envelope in the 660 nm region is characteristic of the Type 1 Cu^{2+}.

Electron paramagnetic resonance and magnetic susceptibility measurements have been essential in detecting the presence of the uniquely different Cu ions bound to the enzyme. Early measurements (57) demonstrated that only 50% of the total Cu was detectible by double integration of the EPR signal, and subsequent

Table 2. Properties of multi-copper oxidases

| Enzyme and source | Laccase | |
	Polyporus versicolor	*Rhus vernicifera*
Chemical properties		
Total number Cu/mole	4 (51)	4 (78, 220, 221)
Hydrodynamic parameters	$s = 4.8$ (50) $\bar{v} = 0.705$ (51)	$s_{20} = 6.25$ s (221) $D_{20,w} = 3.57 \times 10^{-7}$ cm^2 sec^{-1} (221) $\bar{v} = 0.70$ (221) $\bar{v} = 0.734$ (78) $\bar{v} = 0.676$ (214)
Molecular weight	64,400 aa (51) 62,000 S,D (50)	1.04×10^5 (78) 1.10×10^5 (214) 1.41×10^5 (S,D) (221)
Number protein subunits		
Isoelectric point	3.07, 3.27 (Laccase A) (52)	8,55 (214)
Spectral criterion of purity	$A_{280}/A_{610} = 15.2$ (51)	$A_{280}/A_{250} = 2.3$ (214) $A_{280}/A_{614} = 15.2$ (214)
Amino acid composition	+ (51)	+ (214)
Half-cysteine content	6 (51), 7 (55)	6–7 (214); 7 (55)
Sulfhydryl groups	1 (55)	1 (55)
Carbohydrate composition	14% (50, 51)	45% (214)
Disposition of Cu atoms Number Type 1	1	1
Number type 2	1	1
Number diamagnetic	2	2
Oxidation-reduction potentials Type 1	767 mV (63), 785 mV (64)	415 mV (220), 394 mV (64)
Type 2		365 (64)
Type 3	782 mV, $n = 2$ (64)	434 mV, $n = 2$ (64)
Type 3, excess F$^-$	570 mV, $n = 2$ (64)	390 mV, $n = 2$ (64)
Spectroscopic and magnetic properties		
EPR parameters Type 1	Laccase A	(27)
g_{11}	2.19 (59)	2.23
$g_\perp$	2.03	2.05
A_{11} cm^{-1}	0.009	0.0043
Type 2	(59), (65)	(27)
g_{11}	2.24; 1 F$^-$, 2.26; 2 F$^-$, 2.28	2.24
$g_\perp$	2.04	2.05
A_{11}	0.0194 0.0195 0.018	0.020
Optical absorption spectrum	Laccase A 610 nm, $\varepsilon = 4.9$ mM^{-1} cm^{-1} (51) 330 nm, $\Delta\varepsilon = 2.7$ mM^{-1} cm^{-1} (62)	615 nm, $\varepsilon = 5.7$ mM^{-1} cm^{-1} 333 nm, $\Delta\varepsilon = 2.8$ mM^{-1} cm^{-1} (27)
Circular dichroism spectra cm^{-1} or nm	(215) (9) Laccase A Laccase B 13,200 (+) 13,700 (+) 15,800 (−) 16,200 (−) 19,000 (+) 18,800 (+) 22,500 (−) 22,100 (−) 25,500 (+) 28,000 (+)	(215) 11,700 (−) 14,000 (−) 16,800 (+) 19,000 (+) 22,400 (−)

James A. Fee
Copper Proteins — Systems Containing the "Blue" Copper Center
in Structure and Bonding, Volume 23

Errata

Page 15, Table 2, Column "Ceruloplasmin"
8. line from the bottom:

$$610 \text{ nm}, \varepsilon = 6.5 \text{ m}[\text{Cu}]^{-1} \text{ cm}^{-1} \text{ (123)}$$

should read

$$610 \text{ nm}, \varepsilon = 11 \text{ mM}^{-1} \text{ cm}^{-1} \text{ (123)}.$$

Table 2. (continued)

Ceruloplasmin *human serum*	Ascorbate oxidase	
	Green zuchini squash	*Cucumber (191)*
7 (6–8) cf. Table 4	8–12 (*54, 192*) $S_{20,w} = 7.52$ s (*192*)	8 $S_{obsd} = 7.1$ s $\bar{v} = 0.71$
	140,000 S,D (*192*)	132,000 (SE)
1 (*119*) 4.4 (*101*) $A_{280}/A_{610} = 20.8^{-21.9}$	4 (*54*) $A_{280}/A_{610} = 25.6$ (*193*)	6.0–7.8 $A_{280}/A_{607} = 24$
+ (*119*)	+ (Summer crook neck squash) (*189*)	
~15 (*119*) 4 (*126, 55*) 6–10% (*112, 113*)	17–18 (*189*) 10–12 (*189*)	
2 1 4	3 (*225*) 2 4	
490 mV (*123*) 580 mV (*123*)		
cf. Fig. 6	(*225*) 2.227 2.058, 2.036 0.0058	2.22 $g_m = 2.06$ 0.005
(*141*) 2.247 2.06 0.0189	2.242 2.053 0.0199	Present
610 nm, $\varepsilon = 6.5$ m[Cu]$^{-1}$ cm^{-1} (*123*) 330 nm, $\Delta\varepsilon = 3.3$ mM^{-1} cm^{-1} (*119*) 610 nm, $\varepsilon_{1cm}^{1\%} = 0.68$ (*137*)	760 nm (*225*) 610 $\Delta\varepsilon = 1.21$ m[Cu]$^{-1}$ cm^{-1} 330 $\Delta\varepsilon = 0.49$ m[Cu]$^{-1}$ cm^{-1}	760 (*S*) 607 $\varepsilon = 1.2$ m[Cu^{2+}] 330 *S*
(*215*) 13,500 (−) 16,500 (+) 18,700 (+) 22,300 (−)	(*193*) 770 nm (−) 660 (+) 480 (−) 420 (+) 310 (−)	

determinations of room temperature magnetic susceptibility showed only half the total Cu present was paramagnetic (58). A subsequent study (59) revealed the presence of two types of Cu^{2+} having rather different EPR parameters. By using a computational method described by *Vänngård* (60), in conjunction with a variety of chemical experiments, it was possible to show that each ion was present in equal amounts, and these ions were designated Type 1 and Type 2 Cu^{2+} ions (Fig. 3). The four Cu atoms are thus bound in at least three uniquely different environments: Type 1 Cu^{2+}, Type 2 Cu^{2+} ions, and two Cu which are EPR-non-detectable non-paramagnetic at room temperature.

Anaerobic redox titrations (61) have established the presence of approximately four electron accepting sites in the molecule, these being divided among two one-electron acceptors and one two-electron acceptor (62); there are no other functional redox centers in the molecule. The two-electron center has been associated with the 330 nm absorption band (62). The redox potentials of the individual sites, as determined by a combination of various potentiometric and spectrophotometric measurements, are found in Table 2.

Type 1 Cu^{2+}. The Type 1 Cu^{2+} is characterized by the unusually small hyperfine coupling constant in the parallel region of its EPR spectrum. This and its association with the efficient absorption of red light place it in a class with the Cu^{2+} centers discussed in Section II. The optical and magnetic properties of this ion, presented in Table 2, are distinguished only in detail from those of other blue proteins.

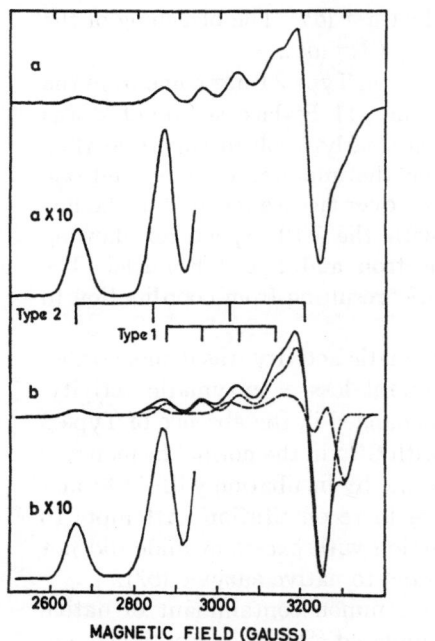 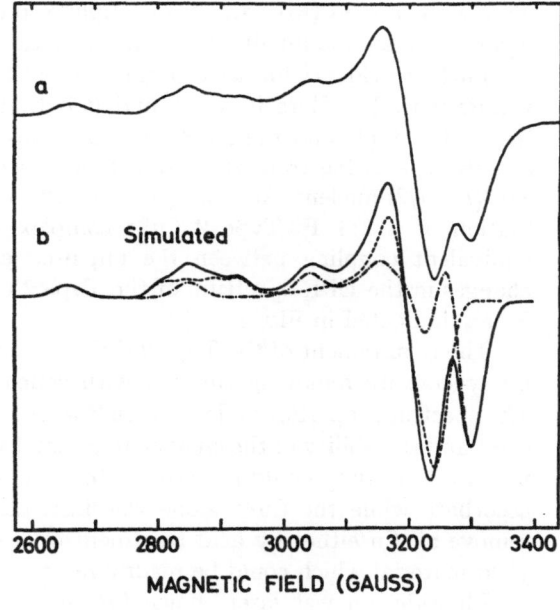

Fig. 3. Experimental and computed EPR spectra of *Polyporous* (left) and *Rhus* (right) laccases. [Taken from *Vänngård*; Ref. (205)]

There is no available evidence that Type 1 Cu^{2+} can be perturbed by external means: it does not bind anions and water does not appear to be a ligand of its first coordination sphere (51). The chemistry of the Type 1 ion is thus dominated by its redox properties.

The redox properties can be divided among those involved with its reduction by potential substrates of the enzyme, its reoxidation by molecular oxygen, and by electron transfer to and from other redox centers within the molecule. Type 1 Cu^{2+} is reduced rapidly by a variety of both one-electron and potential two-electron substrates via a one electron process (63, 64). A likely reason for this ease of reduction of Type 1 Cu^{2+} ion lies with its exceptionally high redox potential (0.77 V at pH 6.2, Table 2). Introduction of molecular oxygen to reduced laccase results in a rapid reappearance of the blue color characteristic of Type 1 Cu^{2+}. The question of intramolecular electron transfer occurring during reoxidation will be considered below.

Type 2 Cu^{2+}. Anions such as azide, cyanide, cyanate, fluoride, and thiocyanate are effective inhibitors of laccase in the concentration range 1—100 μM; while other anions such as phosphate, chloride, nitrate, and sulfate are inhibitory only in the neighborhood of 10 mM (65). To investigate the source of the inhibition, the interaction between the enzyme and fluoride and cyanide was studied by EPR spectroscopy. When CN^- was mixed with laccase the blue color disappeared and the EPR spectrum of the resulting material showed a wealth of superhyperfine structure due to coordinating N-atoms (59, 65). The superhyperfine pattern obtained was complex arising from 3—4 N-atoms (59). When ^{13}CN was used in place of ^{12}CN the superhyperfine pattern was altered and the calculations of Värngård showed two ^{13}CN were coupling to the Cu^{2+} (65). The bleaching of the Type 1 ion is undoubtedly due to its reduction by CN^- (66).

Further evidence for *two* coordination positions on Type 2 Cu^{2+} come from the results with F^-. Here it was found (65) that the 1:1 F^--laccase complex was unusually stable and the EPR spectrum showed clearly resolved superhyperfine structure resulting from the interaction between the unpaired electron and one ^{19}F ($I = 1/2$) nucleus. Addition of an excess of F^- over laccase resulted in the formation of a 2:1 F^--Type 2 Cu^{2+} complex with the EPR spectrum showing equivalent coupling between the unpaired electron and two ^{19}F nuclei. The changes in the EPR spectrum of the Type 2 Cu^{2+} resulting from coordination to F^- are indicated in Fig. 4.

The requirement of the Type 2 Cu^{2+} for enzymatic activity was demonstrated by specifically removing this ion with concomitant loss of enzymatic activity. The spectral properties of Type 1 Cu^{2+} were unchanged in the absence of Type 2 Cu^{2+} and the ability of the enzyme to interact with CN^- in the normal fashion was impaired. Activity could be restored to the enzyme by incubation with Cu^{2+} and ascorbate while the Cu^{2+} alone was ineffective in reconstitution. Attempts to remove all Cu either by acid treatment or reaction with excess cyanide did not yield material which could be readily reconstituted to native enzyme (67).

Fluoride ion was later shown (67) to be a common contaminant of native enzyme and procedures for its removal were formulated (68). The essential step of the procedure involved treatment of the reduced enzyme with mM H_2O_2, and in the course of this study (68) it was found that H_2O_2 produced spectral changes in

17

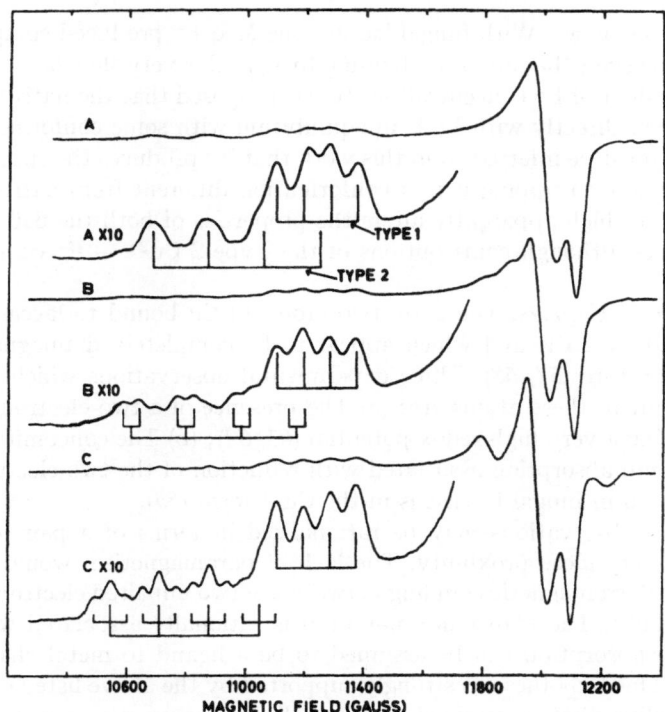

Fig. 4. EPR spectra of *Polyporous* laccase: *A*, the native A form; *B*, plus one molar equivalent F⁻; *C*, plus an excess of fluoride. The spectral parameters of the annated derivatives are listed in Table 2 [Taken from *Malkin, Malmström*, and *Vänngård*; Ref. (*65*)]

the region 320—800 nm. Addition of laccase to 10 mM H_2O_2 resulted in a decrease in 610 nm absorption, the appearance of a new band at 400 nm, and increases in 345 nm absorption. The absorption changes at 345 nm were variable with different preparations and insensitive to previously added F⁻. Over a period of four hours the spectrum had returned to normal. The band at 400 nm, however, was not observed if 1 mM F⁻ was added prior to H_2O_2. The decrease in 610 nm was probably due to reduction of the blue Cu^{2+} by H_2O_2, while the increase in absorption above 500 nm was considered due to a complex formation between laccase Cu^{2+} and H_2O_2. Evidence for the latter came from the 35 GHz EPR spectra which showed that the hyperfine lines of Type 2 Cu^{2+} were shifted to higher field when laccase was treated with excess H_2O_2, and this effect was largely reversed on standing. H_2O_2 was not strongly bound to the oxidized form of the enzyme since a hundred-fold reduction of reagents (from 30 μM enzyme and 100 μM H_2O_2) did not yield any spectral perturbations.

The binding of F⁻ to fungal laccase, tree laccase and ceruloplasmin has been shown to effect substantial perturbations of the entire visible and near ultraviolet spectrum (*69*). While the conditions required for complete binding of F⁻ with the various enzymes are different, the resulting difference spectra are very similar for all three enzymes. With fungal laccase one Meq F^- produced complete optical changes. However, the rate of F^- binding to Type 2 is very slow ($k = 0.0009$ sec^{-1}) and independent of F^- concentration. It was proposed that the native form which does not react directly with F^- is in equilibrium with some conformer which can bind F^-. It must be inferred from this work that F^- produces the spectral changes in the enzyme by trapping it in a conformation different from native enzyme, a conformation which apparently alters the properties of both the 610 and 330 nm chromophores although contributions of the Type 2 Cu^{2+} — F^- complex cannot be excluded.

The Type 3 Coppers. There are two atoms of Cu bound to laccase which are not detected by EPR and which appear to be completely diamagnetic even at room temperature (*57, 58*). Three experimental observations which are thought to be relevant to these atoms are: (a) The presence of a two-electron acceptor in laccase having a very high redox potential (*61, 62*); (b) The concomitant attenuation of 330 nm absorption associated with reduction of the two-electron acceptor (*62*); (c) all Cu in fungal laccase is in the Cu^{2+} form (*55*).

All these observations may be rationalized in terms of a pair of Cu^{2+} ions situated in very close proximity. The lack of paramagnetism would result from a strong antiferromagnetic coupling between the two unpaired electrons associated with each Cu^{2+}. The high redox potential is rationally associated with Cu, and the 330 nm absorption can be assumed to be a ligand to metal charge transfer transition. This hypothesis is strongly supported by the above listed observations. The possibility that a strained disulfide linkage may also account for these properties has been rejected (*61*) on the grounds that, while such a structural unit may have absorption in the near UV region (*70*), it is unlikely that a sufficiently high redox potential could be achieved (*71*). The suggestion of *Byers et al.* (*72*) that the Type 3 Cu consists of a cuprous pair strongly associated with a disulfide bond which breaks on two electron reduction seems excluded by the recent work of *Briving* and *Deinum* (*55*).

The fact that no unusual EPR spectra are observed during titration with reductants or in rapid freeze quench experiments (*61*) suggests that the singly reduced 2ē center does not accumulate to a significant degree and must be quite unstable with respect to both the fully reduced and fully oxidized species. This does not mean that the 1ē reduced site could not form to a small extent, for example, during catalysis.

Results Suggesting Protein Mediated Cu-Cu Interactions. Two types of Cu-Cu interactions must be clearly distinguished:(a) magnetic interactions, for examples, dipole and exchange interactions between the unpaired spins on different Cu^{2+}, the latter must be mediated by intervening ligand atoms, and (b) protein mediated CuCu interaction in which a perturbation at one Cu-site influences the properties of another.

Magnetic Interactions. There will be a dipolar interaction between the two paramagnetic Cu^{2+}, the energy of which will be proportional to 1/(separation distance)^3. However, the fact that the EPR linewidth of the type 2 Cu^{2+} is essentially the same when Type 1 is oxidized or reduced indicating a rather small dipolar interaction and suggesting that the two ions must be separated by at least 8—10 Å in the oxidized protein.

A body of evidence has accumulated which suggests that two Cu^{2+} ions exist as a binuclear cluster in which a strong antiferromagnetic exchange coupling occurs. These ions, Type 3, must be separated by less than 5—6 Å and are probably bridged by electronegative liganding atoms which can mediate the electrostatic exchange interaction (73).

Thus, while these observations have yielded some very rough ideas about the distances between Type 1 and Type 2 and between the EPR non-detectable Cu ions there is no information available concerning the disposition of the Type 3 pair to the other two Cu.

Non-magnetic Interactions. Although CD measurements suggest that the laccase molecule does not undergo a large conformational alternation on either complete reduction or coordination of anions to Type 2 Cu^{2+} (74), a number of observations suggest that non-magnetic interactions occur which are mediated by the protein. Protein mediated Cu-Cu interaction is evident in the effect of binding a single F^- to the Type 2 Cu^{2+} on the spectral properties of the other Cu^{2+} centers, the effect of excess fluoride on the redox potential of the two-electron acceptor, and on the rate of reduction of the two-electron acceptor in the presence of substrates.

Brandén et al. (69) assert that F^- binding to Type 2 Cu^{2+}, as measured by spectral changes at 320, 380, and 600 nm and by an F^--specific electrode, requires approximately one hour to achieve complete binding. The rate limiting step in F^- binding involves a conformational change of the enzyme prior to F^- binding which has a rate constant of $9 \cdot 10^{-4}$ sec $^{-1}$ ($t_{1/2} = 13$ min). While the reaction rate seems to be inconsistent with the experiments of *Malkin et al.* (65), the kinetics require that F^- binds to Type 2 Cu^{2+} in the following reaction sequence:

$$E \underset{k_1}{\overset{k_1}{\rightleftarrows}} E'$$

$$E' + F^- \xrightarrow{k_2} E' \cdot F$$

where k_1 is rate limiting and E′ F has altered spectral properties.

The spectral changes in the region of 600 nm must be associated with the Type 1 Cu^{2+}, but *Brandén et al.* (69) question whether the decrease in 600 nm absorption is due to a perturbation of this ion or only a partial reduction by endogenous substrate (12). However, the decrease in 380 nm and increased 320 nm absorbance are interpreted as a perturbation of the Type 3 Cu chromophore. Regardless of the interpretation of certain spectral changes F^- binding to Type 2 Cu^{2+} clearly influences the individual chromophores of laccase.

Very strong support for protein mediated Cu-Cu interactions comes from observations concerning the effect of F^- binding on the redox behavior of the Type 1 ion and the two-electron acceptor. During the reductive titration of laccase in the absence of F^- the 330 nm and 610 nm bleaching occurred simultaneously, indicating the several electron accepting sites to have similar redox potentials (61). However, in the presence of excess F^- the Type 1 ion was completely reduced before any attenuation of 330 nm absorption (62). *Reinhammer* (64) later determined the redox potential of the Type 3 center in native and F^- treated enzyme to be 0.78V and 0.57V, respectively. These results suggest that the Type 3 center is sensitive to the state of ligation of the Type 2 Cu^{2+}. As might be expected, F^- bound to Type 2 Cu^{2+} also has effects on the catalytic properties of the enzyme (see below).

A further example of apparent protein mediated interaction between the Cu binding sites involves the autoreduction of the enzyme at high pH. Above approximately pH 5 laccase undergoes a reversible bleaching of its blue color. This transition is adequately described by the ionization of a single conjugate acid having a pK_{app} of approximately 7.4 (75). Concomitant with the loss of blue color there is a reduction of the Type 1 Cu^{2+} ion while the Type 2 and diamagnetic centers apparently are unchanged. This was indicated by the fact that only $2\bar{e}$ can be accepted by the protein at high pH, and the difference spectrum, laccase at pH 5 minus laccase at pH 7.8 showed absorbance differences in the region of the 610 nm band but not in the 330 nm region (12). Finally, the EPR spectrum at high pH closely resembled that arising from Type 2 Cu^{2+} (12, 75).

The nature of the reductant is not known although enzyme preparations must contain a considerable excess of an endogenous source of electrons (12). In fact, in the presence of oxygen the enzyme is slowly but continually oxidizing this substance, presumably by the usual oxidative mechanism which is known to involve reduction of Type 1 Cu^{2+} (12). Thus, the enzyme will be reduced to a degree determined by the relative rates of reduction and oxidation. Since these two processes depend on the concentrations of oxygen and reductant no structural deductions can be made from the apparent pK_a of the Type 1 reduction. The Type 1 ion could thus be reduced more rapidly or oxidized more slowly at high pH. The simplest interpretation of these results is that above pH 5.5 the enzyme has a lower rate of reoxidizing Type 1 Cu^+ and if, as has been suggested (76), reoxidation of Type 1 ion involves intramolecular transfer from Type 1 Cu^+ to the two-electron center, then this process may be slowed at high pH. Some support of this idea comes from the fact that Type 1 ion reduced at high pH is rapidly reoxidized under anaerobic conditions upon adjusting the solution to low pH (12). Presumably this represents electron transfer to another accepting site in the molecule.

2. _Rhus vernicifera_ Laccase. The physical and chemical properties of Rhus laccase and fungal laccase may be compared in Table 2. It is clear that while there are differences in the gross physical properties such as molecular weight, carbohydrate content, and amino acid composition, nature has preserved a four-Cu complex capable of accepting four electrons (77—79) consisting of one Type 1 Cu^{2+}, and one Type 2 Cu^{2+}, and the Type 3 pair. There are other minor differences reflected in the comparable redox potentials, g values, and other detailed spectroscopic parameters but on the whole the state and function of Cu bound to _Rhus_ laccase would seem to be identical with _Polyporus_ laccase (cf. Fig. 3).

The sulfhydryl chemistry of _Rhus_ laccase differs slightly from that of fungal laccase but has considerable bearing on all investigations concerned with the ligand environment of the Type 1 Cu^{2+}. _Briving_ and _Deinum_ (55) have shown that native _Rhus_ laccase has a single freely reactive SH group. When this residue is combined with a mercurial the native properties of the enzyme are unchanged, and when denatured under anaerobic conditions in the presence of excess EDTA, the protein demonstrates only one SH group and 2 disulfide entities. Therefore, the Type 1 Cu^{2+} does not appear to gain any of its unusual properties by close association with the cysteine sulfhydryl group.

There are two notable differences between these proteins: (a) *Rhus* laccase but not the fungal enzyme can be readily reconstituted from an apoprotein and Cu(I) *(81—84)*. This process has not been studied in detail using physical methods to examine the Cu binding during the reconstitution process. (b) *Makino* and *Ogura (63)* have observed that hydrogen peroxide will oxidize only the Type 3 Cu of fully reduced enzyme and *not* the Type 1 Cu^+. In contrast hydrogen peroxide can remove electrons from all the Cu centers of the reduced fungal enzyme *(J. A. Fee* and *B. G. Malmström*, unpublished observations), but the rate of reoxidation of the Type 1 Cu^+ is approximately 10^5 times slower than that observed with molecular oxygen [Ref. *(94)* and *I. Pecht*, unpublished observations].

B. The Mechanism of Laccases

The stoichiometry of the laccase catalyzed reaction *(77, 42, 80)* has long been established as:

$$2\,AH_2 + O_2 \longrightarrow 2A + 2H_2O\,.$$

With the exception of a study carried out with a partially characterized multi-copper oxidase isolated from tea leaves *(85)*, there has been very little detailed work concerned with the steady state kinetic behavior of laccases. Early work on the transient kinetics indicated, however, that: (1) enzyme bound Cu^{2+} was reduced by substrate and reoxidized by O_2, and (2) substrate was oxidized in one-electron steps to give an intermediate free radical in the case of the two electron donating substrates such as quinol and ascorbic acid. The evidence obtained suggested that free radicals decayed via a non-enzymatic disproportionation reaction rather than by a further reduction of the enzyme *(86—88)*. In the case of substrates such as ferrocyanide only one electron can be donated to the enzyme from each substrate molecule. It was clear then that the enzyme was acting to couple the one-electron oxidation of substrate to the four-electron reduction of oxygen via redox cycles involving Cu.

There has been a good deal of discussion as to how this is achieved *(61, 56, 89, 76, 90, 91)*; two mechanistic extremes are: (1) The reduction of oxygen by discrete one electron steps with the enzyme serving to stabilize the intermediate species (superoxide, peroxide, and hydroxyl) formed during this process. (2) The simultaneous (or nearly so) reduction of oxygen by an event involving transfer of four electrons to molecular oxygen. In this case the enzyme would serve much as a capacitor, storing four electrons and subsequently discharging them to oxygen.

The finding that the redox centers of fungal lacase have reduction potentials very near that of the O_2/H_2O couple [Table 2, *(92)*] raised an interesting question concerning the high rates of reoxidation of reduced enzyme by molecular oxygen. Taking a very simple approach to the energetics of the reoxidation process it can be reasonably demonstrated that substantial energetic barriers must be overcome if either one-electron or two-electron schemes are operative. Any usefulness of the following arguments would extend to all the multicopper oxidases.

Considering the information presented in Table 3 and disregarding for the moment effects of binding the putative radicals to the enzyme; it is clear that the only favorable steps in the reoxidation are the reduction of O_2 to H_2O_2, the two-electron reduction of H_2O_2 to H_2O, and the four-electron reduction of O_2 to $2H_2O$. The one-electron oxidation of Type 1 by O_2 is very unfavorable and would require unusual stabilization of the formed O_2^-. These data suggest that a multielectron transfer from a reduced form of laccase to oxygen could overcome these energetic barriers: (a) A two-electron reoxidation of the enzyme by the O_2/H_2O_2 couple is unfavorable but this barrier could be overcome by moderately strong binding of H_2O_2 to the enzyme and two-electron reoxidation by the H_2O_2/H_2O couple is very favorable, there being no necessity to stabilize any intermediates. (b) A one-electron mechanism is unfavorable to the level of ·OH and strongly favorable from ·OH to H_2O. A general conclusion which can be made from these considerations is that either four electrons are simultaneously transferred to oxygen or the intermediate reduction products of oxygen are quite strongly stabilized by binding to the enzyme. *Pecht (93)* has shown that pools of isotopically distinguishable H_2O_2 and O_2 remain unmixed by the action of the enzyme. Thus, if H_2O_2 is formed during the enzymatic action it is not released from the enzyme prior to being reduced to H_2O. A further general conclusion which can be made is that these enzymes must have a unique oxygen binding site capable of efficiently orchestrating the reduction of oxygen. It is reasonable to associate the Type 3 Cu centers with this site.

Table 3. Reduction potentials and free energies of possible reactions between the copper centers of *polyporous* laccase and some species of reduced oxygen

Reaction	ΔE^{o1} (Volt)[a]	ΔG^{o1} (Kcal/mole)
O_2 + Type 1 Cu^+ $\rightleftharpoons$ O_2^- + Type 1 Cu^{2+}	− 1.36	31.0
O_2 + 2 Type 3 Cu^+ + 2H^+ $\rightleftharpoons$ H_2O_2 + 2 Type 3 Cu^{2+}	− 0.3	13.8
O_2^- + Type 1 Cu^+ + 2H^+ $\rightleftharpoons$ H_2O_2 + Type 1 Cu^{2+}	0.21	− 4.8
H_2O_2 + H^+ + Type 1 Cu^+ $\rightleftharpoons$ ·OH + H_2O + Type 1 Cu^{2+}	− 0.39	9.0
H_2O_2 + 2H^+ + 2 Type 3 Cu^+ $\rightleftharpoons$ 2H_2O + 2 Type 3 Cu^{2+}	1.56	− 35.9

[a]) The reduction potentials for the Cu sites were taken from Table 2 and the reduction potentials for the oxygen species from Ref. *(92)*.

In view of the energetic considerations a good deal of effort has been expended in determining the rates of reduction and oxidation of the Type 1 Cu center and the diamagnetic CuCu pair. The relevant observations for the two laccases are recounted below.

1. ***Polyporus versicolor* Laccase.** Under all studied conditions the Type 1 Cu^{2+} is rapidly reduced by substrates. With ferrocyanide the reaction follows a second-order law during the initial phase when substrate concentration is below

K_m, but it is too fast to observe at higher concentrations. The Type 1 center has been considered for some time as the initial point at which electrons from substrate enter the laccase molecule (94, 95). In the absence of oxygen, the reduction of the Type 3 Cu-pair is unimolecular at high substrate concentration and is very slow ($k = 1-2$ sec^{-1}). Type 3 reduction is also independent of the nature and concentration of substrate and of enzyme (62, 90, 95). It has been proposed that this slow reduction results from an internal oxidation of Type 1 Cu^{2+} by Type 3 Cu (90, 95). Fluoride ion strongly inhibits the reduction of Type 3 Cu ($k = 0.008$ sec^{-1}) (95), but does not change the qualitative behavior of the reaction. The important fact is that whether fluoride is present or absent the reduction as observed by transient kinetics occurs much too slowly to be a viable step in the catalytic action (62, 90, 94).

The levels of reduction of the Type 1 and Type 3 Cu ions have been measured in the steady state (62). Under conditions where oxygen is not limiting the Type 3 pair, as measured by the 330 nm absorption, remains fully oxidized while the Type 1 ion is reduced to a degree dependent on the nature and concentration of substrate. The approach to steady state conditions have also been studied by observing both reduction of Type 1 ion and product formation (94). Here the Type 1 ion is reduced beyond the steady state level depending on substrate concentration (cf. Fig. 5). Product formation, as observed with ferricyanide, follows a second order law during the very early stages of reaction, and before entering the zero-order condition there is a brief period in which product formation is slower than that in the steady state. This is particularly evident in Figs. 4 and 5 of Ref. (94) where the S/E ratios were 15 and 12, respectively. This lag cannot be due to a negative contribution of absorbance due to reduction of Type 3 Cu since oxygen was not limiting in these experiments thus no reduction of the Type 3 ions would be expected to occur (62).

The approach to steady state from enzyme reduced by an excess of quinol has also been reported, however, only the Type 1 ion was observed (94). In this experiment there was an overshoot in the reoxidation of Type 1 ion which was followed by a reduction to the steady state level as long as the initial concentration of oxygen was greater than the excess concentration of quinol.

Thus, two situations have been recorded in which the enzyme system, upon approaching the steady state, appears to exceed that condition with subsequent "relaxation" to the steady state: (a) reduction by excess substrate in the presence of oxygen, and (b) reoxidation by O_2 in the presence of excess substrate. This behavior must account for the small lag in product formation mentioned above, and these observations are consistent with a scheme in which the protein assumes a conformation during the steady state that is different from both the fully oxidized and reduced conformations. It must be kept in mind that the individual rates of reduction and oxidation of the various redox sites within the enzyme might differ among the steady-state and ground- or end-state conformations (76).

The reoxidation of fully reduced enzyme by O_2 has been shown to be rapid (94) but with partially reduced enzyme reoxidation is considerably slowed and an intermediate component is formed which absorbs at 420 nm (96). The intermediate substance is thought to be a complex of Type 2 Cu^{2+} and H_2O_2 since similar

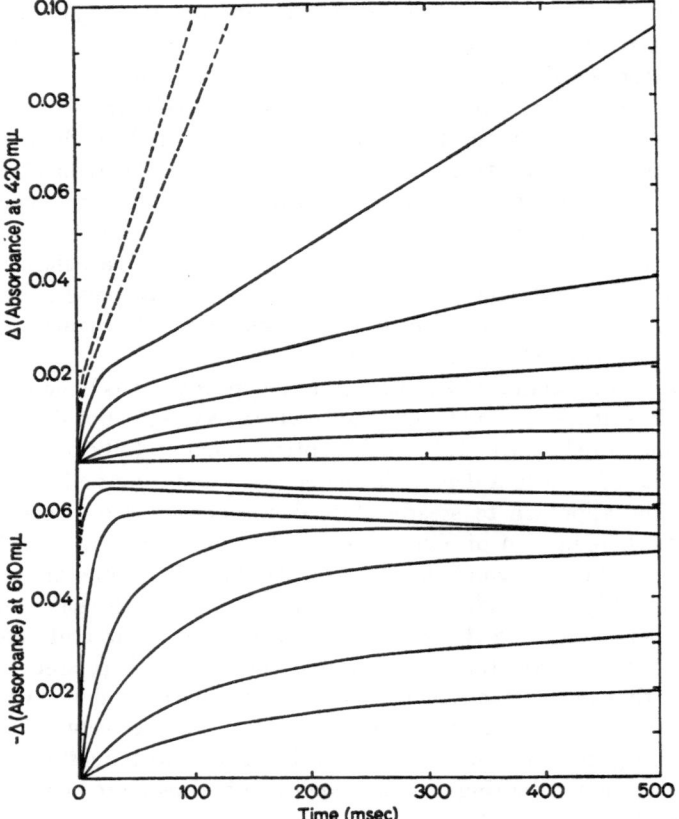

Fig. 5. Time course of the approach to the steady state in the system: laccase, ferrocyanide, and oxygen at pH 5.4. The reduction in laccase (bottom frame) and ferricyanide formation (upper frame). The concentration of laccase was 8.3 μM, the concentration of ferrocyanide was 2.5, 5.0, 12.5, 25, 125, 500 and 1000 μM and the concentration of oxygen was 275 μM. Note that with the higher concentrations of ferrocyanide Type I Cu^{2+} is initially reduced beyond the steady state level and significant reoxidation occurs during the steady state. Also note the distinct lag in product formation subsequent to the initial burst and prior to the zero-order condition. This is most evident in the traces corresponding to 25 and 50 μM ferrocyanide. [Taken from *Malmström, Finazzi-Agro,* and *Antonino*; Ref. (*94*)]

spectral changes can be induced in the native enzyme by adding H_2O_2 (*68*). The spectral changes which occurred in the latter experiment are inhibited by F^-, presumably by competition for Type 2 Cu^{2+}. The relatively slow reoxidation to the fully oxidized form appears to be due to the limiting decomposition of the Cu^{2+}—H_2O_2 complex.

It was noted earlier that H_2O_2 does not bind very strongly to the fully oxidized protein (*68*). Clearly, if H_2O_2 is formed during the course of oxidative activity it must be bound much more strongly than to oxidized protein. Of course, this is possible and one example of the dependence of affinity on the valence state of the various redox centers concerns the binding of F^-. During the initial phase of the catalytic reaction one Meq of F^- effects substantial inhibition but as catalysis proceeds this is overcome (69). Apparently an intermediate form of the enzyme accumulates which has a reduced affinity for F^-. Such behavior is consistent with the concept that the behavior of the system is strongly influenced by protein mediated Cu-Cu interactions.

2. *Rhus vernicifera* Laccase. *Holwerda* and *Gray* (97) have recently published an extensive study on the mechanism of reduction of Rhus laccase by hydroquinone. This work holds many implications for the direction of research in multicopper oxidases as it showed: (1) that there are significant differences between the kinetic behavior of *Rhus* and *Polyporus* laccases; (2) that the individual rate constants may be strongly dependent on ionic strength, temperature, and pH; and (3) that the general pattern of reactivity of these systems may be extremely sensitive to different combinations of environmental conditions. Specifically, their major findings are: (a) In contrast to fungal laccase, both Type 1 and Type 3 centers are reduced by second order processes apparently involving HQ^- (the hydroquinone anion) as the specific reductant. (b) Fluoride ion, assumed to be bound to Type 2 Cu^{2+} (98), inhibits the reduction of both Type 1 and Type 3 centers while with Polyporus laccase F^- binding affects only the rate of reduction of Type 3 Cu (95). (c) In the presence of azide, Type 1 Cu^{2+} is reduced in a slow, substrate independent, process while the attenuation of the 405 nm band (due to $N_3^- \rightarrow Cu^{2+}$ charge transfer transition) was reduced even more slowly by a process which showed an extraordinary first order dependence on azide concentration.

Holwerda and *Gray* (97) proposed a mechanism for the reduction process involving a central role of Type 2 Cu^{2+} as the initial point at which electrons enter and are subsequently distributed to the other electron acceptors. This interpretation would seem to be supported by the very recent observation of *Brandén* and *Reinhammar* (98) that the Type 2 ion of Polyporus laccase is reduced and subsequently reoxidized in a very short time period. The authors also emphasize the parallel behavior of the Type 1 and Type 3 centers; it is particularly striking that under a variety of conditions the rates of Type 1 and 3 reductions are very similar. Indeed, when Cr^{2+} was used as the reductant, a similar observation was made (99). Perhaps Cr^{2+} reduces the enzyme via a bridging ligand (H_2O?) between it and Type 2 Cu^{2+}.

It is obvious from the above discussion that the existing number of observations is insufficient to allow the formulation of a general catalytic reaction scheme for these enzymes. In spite of the differences in the presteady state behavior of fungal and *Rhus* laccases, it seems unlikely that they utilize different catalytic mechanisms, and the two enzymes probably behave identically during steady state catalysis. Elucidation of the mechanism will thus require knowledge of their steady state structure and behavior.

C. Ceruloplasmin

In 1944 *Holmberg* (100) observed that a particular serum protein fraction which was bluish in color possessed oxidase activity toward paraphenylene diamine, paracresol, and catechol. The blue protein was subsequently purified by *Holmberg* and *Laurell* (*101, 102*) and given the name *ceruloplasmin*. These workers made a number of fundamental observations on the physical and chemical properties of ceruloplasmin the significance of which have only recently been appreciated, and these will be recounted in the body of this discussion.

The function of ceruloplasmin in the mammalian organism is presently unknown although there have been suggestions that it is required for the efficient incorporation of iron into transferrin (*103*), and that it is an essential component of copper mobilization (*104—106*). These have been the subject of considerable debate (*107, 108* and references therein), and will not be of further concern here.

Normal human serum actually contains two chromatographically distinct forms of ceruloplasmin. A calcium phosphate column cleanly separates major (I) and minor (II) components, the latter being approximately 15% of the total ceruloplasmin present in the serum (*109—111*). *Ryden* (*112, 113*) has shown that the two forms though identical in amino acid composition and Cu binding capacity, differ in the degree to which carbohydrate components have been affixed to the protein during biosynthesis. Most of the presently accepted physical properties of human ceruloplasmin are presented in Table 2.

The protein has a molecular weight near 134,000 daltons, consists of a single polypeptide chain composed of about 1050 amino acid residues, and has 34—36 associated carbohydrate residues. The complex nature of this molecule has resulted in a number of confusing results. Notably, with respect to the molecular weight, the number of protein subunits, and the number of Cu atoms bound per molecule of protein.

There are several reports in the literature demonstrating that ceruloplasmin is made up of multiple peptide chains (*114—118, 226*). *Rydén* (*119*) recognized that ceruloplasmin was extremely sensitive to proteolytic cleavage and broached the question of whether the "sub units" of ceruloplasmin may not arise from degradation by endogenous proteases. By introducing a general protease inhibitor, ε-aminocaproic acid, at the beginning of the enzyme preparation he (*120*) was able to isolate electrophoretically pure ceruloplasmin from human, pig, rabbit, and horse (*121*) which consisted of a single polypeptide chain (*122*). *Rydén*'s conclusion has been questioned by *Freeman* and *Daniel* (*226*) who have claimed that longer times are required for dissociation in SDS and reducing agents than was realized by *Rydén*. While this factor may be important, it should be recognized that *Freeman* and *Daniel* carried out their experiments with a commercially prepared sample of human ceruloplasmin whose history was not delineated. This compromises the conclusions of these authors as *Rydén* has argued that the protease inhibitor must be introduced at the start of the fractionation procedure, not at some later time. It is important to emphasize at the outset that the peptide bond cleavage experienced during the earlier preparative procedures leaves the physical properties of the bound Cu ions and the catalytic capability of the human enzyme (*123*) largely unaltered. However, previous results concerned with the quarternary structure of the enzyme would seem to be generally superceded by the work of *Rydén*.

Ceruloplasmin appears to have an elongated shape as *Kasper* and *Deutsch* (*124*) report an axial ratio of 11 based on sedimentation and diffusion studies. *Hibino* and *Samejima* (*125*), on the basis of ultraviolet circular dichroism, and infrared absorption measurements have suggested that porcine ceruloplasmin is composed primarily of β- and random coil types of tertiary structure. Since, the physical properties of the Cu²⁺ ions are not altered by the limited proteolysis which occurs during isolation, it is probable that gross conformational alterations likewise do not occur and the above conclusions would appear appropriately valid for uncleaved enzyme.

1. Chemistry of the Individual Amino Acids and Carbohydrate Residues. *Sulfur containing amino acids.* Ceruloplasmin contains 17—18 methionine residues and 14—15 half-cysteine residues *(119, 124)*. *Kasper* and *Deutsch (124)* reported that ceruloplasmin has a single —SH group which is readily alkylated in both holo and apo- forms of the protein.

Witwicki and *Zakrzewski (126)* have extended the observation, finding that 1.00 ± 0.03 sulfhydryl per 150,000 daltons were available in 0.7 M Tris HCl for reaction with the sulfhydryl reagent, β-hydroxyethyl-2,4-dinitrophenyl disulfide. In the presence of 8 M urea they observed 3 reactive SH groups, and in the presence of both urea and mM EDTA 3.6 SH groups were observed to react. The quantitative reactivity of the SH group in native protein was consistent in the presence of high concentrations (~0.7 M) Tris HCl while phosphate buffers produced an erratic behavior. *Rydén (119)* has related the data of *Witwicki* and *Zakrzewski (126)* to a lower molecular weight assumption obtaining 0.97 available SH in the native protein and 2.85 and 3.3 in the unfolded protein in the absence and presence, respectively of EDTA. All half-cysteine was thus accounted for as follows: 1 freely reacting SH, 3 buried SH, and 12 cysteine associated as disulfide linkages. By difference, the native ceruloplasmin molecule may contain six disulfide bridges. *Witwicki* and *Zakrézewski (126)* have also isolated the ceruloplasmin derivative in which the free SH group is blocked in a disulfide combination with —S—CH₂CH₂OH, and report that it has full enzymatic activity and an unaltered visible optical spectrum. Apparently the free SH is in no way associated with the active sites of the enzyme.

Very recently *Briving* and *Deinum (55)* have shown that ceruloplasmin prepared from fresh serum, in the presence of ε-amino caproic acid, has no freely reacting sulfhydryl groups. These authors have confirmed that fully denatured ceruloplasmin expresses four SH groups with the remainder of cysteine sulfur existing as disulfide bridges. Solutions of ceruloplasmin which become slightly denatured (not inactivated) on standing express a single reactive SH group *(55)* which is not critical to the functioning of the native enzyme as was observed previously *(126)*.

N-terminal Amino Acid

Kasper and *Deutsch (124)* report 0.9 residues of valine and 0.3 residues of lysine per molecular weight 160,000. Later *Ryden (122)*, working with proteolytically unaltered protein observed only a single valine per 134,000 g protein.

Carboxylic Amino Acid Residues

Based on the similarity of the proton titration behavior of apo- and holo-ceruloplasmin *Kasper* and *Deutsch (124)* concluded that carboxyl groups did not coordinate to the copper.

Tyrosine

Kasper and *Deutsch* (*124*) have titrated the approximately 65 tyrosine residues spectrophotometrically. The titration between pH 9 and 12 was not reversible for either native or apo-ceruloplasmin but the magnitude of the spectral changes and the degree of hysteresis observed on back titration from pH 13 were substantially the same with native apo forms. No indication was obtained that tyrosine and residues were either involved in the binding of Cu or participated in any unique structural feature of the enzyme.

Histidine

Bannister and *Wood* (*127*) submitted ceruloplasmin to methylene blue mediated photo-oxidation, and were able to correlate loss of oxidase activity with the destruction of histidine residues and the loss of Cu from the enzyme. The one-to-one correspondence between histidine loss and oxidase activity led these authors to conclude that the active region of the molecule contains an essential histidine residue. *Nylén* and *Pettersson* (*128*) have criticized this conclusion because the non-specific nature of the photo-oxidation [cf. Ref. (*128*)] was not taken into account. The latter authors found that diethylpyrocarbonate reacted specifically with the histidine residues at pH 6. Approximately 47 histidine residues reacted in 6 M urea, agreeing quite well with the 43 histidine residues detected by amino acid analysis (*124*). When approximately 50% of the total histidines had been carbethoxylated in the native oxidized form, no spectral perturbation of the enzyme had occurred. However, beyond 10% carbethoxylation activity decreased linearly and extrapolated to zero at 100% reaction indicating that all imidazole groups must be modified for full loss of activity. Hydroxylamine could reactivate all but the most extensively carboxethylated samples. Integration of the EPR spectra of 50% carboxethylated samples showed no change in total intensity, and that the only change was a slight modification of the Type 2 EPR signal (see below) which could be correlated with loss of activity. The authors conclude that while all histidines must be reacted for loss of activity not all histidine residues contribute to the activity, and clearly modification of these residues does not modify the Cu binding sites.

Carbohydrate Residues

Jamieson (*129*) demonstrated that each molecule of ceruloplasmin possesses 9 to 10 carbohydrate chains consisting of sialic acid, galactose, mannose, fucose, and N-acetyl glucosamine, and he has also shown that sialic acid could be removed from these glycopeptides by neuraminidase. [*Rydén* (*119*) reports 12.1 residues of galactose per molecule of ceruloplasmin.] This derivative has been used to explore the physiological behavior of ceruloplasmin (*131, 132*). *Morell et al.* (*130*) have successfully removed all sialic acid from native ceruloplasmin by incubation with neuraminidase for a period of approximately 4 days. Asialoceruloplasmin was shown to have physical properties substantially identical with native protein and to have lost no Cu during the treatment with neuraminidase.

Sialic acid is attached to galactose in a preponderance of glycoproteins and ceruloplasmin does not differ in this respect. Removal of the sialic acid exposes a galactose residue, and *Morell et al. (130)* demonstrated that the terminal galactose residues of asialoceruloplasmin could be oxidized by galactose oxidase and subsequently reduced with tritium labelled sodium borohydride. Approximately four of the 9—10 galactose residues were titrated in this fashion.

2. Molecular Weight and Copper Content. Since its initial characterization by Holmberg and Laurel, ceruloplasmin was thought to have a molecular weight of 150,000—160,000, and all early measurements of protein and copper were related to this value, However, in 1969 *Magdoff-Fairchild et al. (133)* carefully assessed the molecular weight of human ceruloplasmin in single crystals to be 132,000 $\pm$ 4,100 from a crystallographic measurement of the unit cell volume, the density of the crystals, and careful determination of the partial specific volume of the protein *(134)*. Later, *Rydén (122)* reported a molecular weight of 134,000 and *Freeman* and *Daniel (226)* one of 124,000 both being based on sedimentation-equilibrium studies, and in addition, the amino acid and carbohydrate composition suggest a molecular weight near 130,000 *(119)*. These studies appear to establish the molecular weight of ceruloplasmin to be near 124,000—134,000. *Morell et al. (134)* have carefully determined the Cu content and dry weight of the protein finding a value of 0.275 $\pm$ 0.009%, w/w Cu which for a molecular weight of 132,000 $\pm$ 4,100 corresponds to 5.73 $\pm$ 0.36 Cu per mole ceruloplasmin. This number is reasonably rounded to 6 Cu per mole protein.

However, the presence of 6 Cu atoms requires that the magnetic properties of Cu bound to ceruloplasmin be rationalized in a presently unacceptable manner. The essence of this dilemma is that while all Cu bound to ceruloplasmin is formally Cu^{2+} *(55)* only 44% of this Cu is paramagnetic *(58, 135)*. This would correspond to 2.6 Cu^{2+} and 3.4 non-paramagnetic Cu atoms, numbers which deviate too greatly from the integer values required by the magnetic properties.

Earlier studies by *Holmberg* and *Laurell (101)*, by *Deutsch* and co-workers *(136, 137)*, and *Nakagawa (138)* on the Cu content of ceruloplasmin indicated ∼0.32% Cu which, based on a molecular weight of 132,000 indicates the presence of 6.65 Cu which reasonably rounds to 7 Cu per protein molecule. 44% of this corresponds to three paramagnetic Cu^{2+} and four non-paramagnetic Cu. While this result is more satisfying from the point of view of the paramagnetic measurements, the discrepancy in the analytical results cannot be lightly dismissed. It is not clear from the literature whether the different Cu contents of the various preparations result from the different analytical procedures used or perhaps even to regional differences in the Cu content of ceruloplasmin fractions. A summary of various analytical determinations is given in Table 4.

We turn now to the physical properties of the Cu bound to ceruloplasmin. The EPR and visible optical spectra suggest considerable similarity between the Cu of ceruloplasmin and that of laccases. Thus, as first reported by *Holmberg* and *Laurell (101)* the optical spectrum is characterized by absorption bands at 610 nm and 330 nm. The EPR spectrum shows the presence of both Type 1 and Type 2 Cu^{2+} ions and a significant fraction of the Cu atoms are EPR-nondectable, 56% as compared to 50% in laccase *(58, 135)*. Thus, there appear to be only three general types of Cu binding sites in ceruloplasmin, and the number of each type of Cu has recently been determined with considerable certainty.

Table 4. Molecular weight and copper determinations of ceruloplasmin

Molecular weight	Method of determination	Purity		% Cu	g-atom Cu 132,000	Ref.	Year
		A_{280}/A_{610}	Cu/gN				
151,000	S,D			0.32	6.66	(101)	1948
		24.0	0.0207	0.31–0.32	6.4[a]	(195)	1958
		21.5		0.34	7.06	(136)	1960
		22.7	0.0197–0.0203		6.14–6.33[a]	(199)	1960
160,000	S,D	21.9	0.0216	0.31–0.32	6.45–6.66	(137, 124)	1962
160,000	S,D	22	0.0216	0.32	6.66	(124)	1963
148,000	S,D	31.4	0.0210	0.34	7.1	(116)	1966
155,000	(Archibald)						
143,000	S,D	24.4		0.30	6.2	(115)	1966
		21.9		0.276	5.72	(130)	1966
				0.27–0.288	5.6–6.0	(200)	1967
132,000	Crystallography	21.7	0.0183[a]	0.275	5.73 ± 0.36	(133, 134)	1969
160,000	Light scattering			0.31	6.4	(138)	1972
134,000	Sed. equil.	22.2				(122)	1972
124,000	Sed equil.	22.0				(226)	1973

[a]) Assuming 15% nitrogen [cf. (124, 137)].

3. The State of Copper in Ceruloplasmin. The early room temperature magnetic susceptibility studies of *Ehrenberg et al.* (58) showed that only about 40% of the Cu bound to ceruloplasmin was paramagnetic. *Aisen et al.* (135) have carefully measured the paramagnetic contribution to the magnetic susceptibility of ceruloplasmin over the temperature range 2—4.2 °K. In this temperature range Curie's law is followed precisely, and the total susceptibility is exactly proportional up to a measuring field strength of 18 kgauss. These authors find that only 44 ± 2% of the total Cu is paramagnetic, *i.e.*, exists as magnetically dilute Cu^{2+}. Both the low temperature and high temperature magnetic susceptibility results are in essential agreement that approximately 40% of the Cu is paramagnetic, and this result is consistent with values obtained by integration of the EPR spectrum of ceruloplasmin which indicated 43—48% Cu detectable as Cu^{2+} in the earliest study (57) and 43% in a more recent work (123). Unless all ceruloplasmin preparations possess a yet undetected heterogeneity in the occupation of the Cu binding sites the magnetic susceptibility and EPR integrations obviate the presence of an even integral number of Cu^{2+} ions bound to each protein molecule. Thus, both 6 and 8 atom Cu per mole protein would seem to be unlikely possibilities.

There are conflicting reports in the literature on the ratio of the number of Type 1 to Type 2 Cu^{2+}. These are based on deductions from EPR integrations and spectral simulations and from chemical studies. *Vänngård, (60)* who pointed out the possible importance of the Type 2 Cu^{2+} in the multicopper oxidases and described a computational procedure for estimating its quantitative presence, reported that the fractional contribution of this distinct Cu^{2+} ion varied with the history of the preparation. Thus in crystallized samples prepared by the method of *Deutsch et al. (137)* having integrated EPR intensity corresponding to 39% of the total Cu^{2+} the Type 2 signal corresponded to 34% of the total EPR signal (13.3% of total copper). With ceruloplasmin freshly prepared from serum the latter percentage was 25%, and with protein from Cohn fractions the intensity of the Type 2 signal was roughly equal to that of the Type 1 signal. *Andréasson* and *Vänngård (139)*, working with commercial samples of ceruloplasmin found a ratio of one Type 2 to Type 1 Cu^{2+} were present in ceruloplasmin. *Veldsema* and *Van Gelder (140)* suggested on the basis of anaerobic reductive titrations that the ratio of Type 2 to Type 1 Cu^{2+} was 3 but recognized that a lower value was also possible.

Cognizant of the considerable difficulties associated with sample history, *Deinum* and *Vänngård (123)* reinvestigated the problem using a variety of ceruloplasmin samples including those prepared from fresh serum in the presence of ε-aminocaproic acid to avoid proteolytic modifications *(120)*. The EPR detectable Cu^{2+}, corresponded to 43% of the total Cu in their preparations. The total EPR signal was comprised of 33% of Type 2 Cu^{2+} and 67% Type 1 Cu^{2+}. Thus, the best preparations of ceruloplasmin appear to contain Type 1 and Type 2 Cu^{2+} in a ratio of 2:1. In this definitive work it was also shown that there are two different Type 1 ions having different optical extinctions and redox potentials. *Gunnarsson et al. (141)* have confirmed that the ratio of Type 1 to Type 2 Cu^{2+} is 2 and have suggested that one of the Type 1 centers in human ceruloplasmin has a slightly larger hyperfine coupling constant (A_{II}) than the other, and that the apparent A_{II} values of one center changes slightly between pH 7 and 5.5 (cf. Fig. 6).

The original suggestion *(67)* that the diamagnetic Cu of multicopper oxidases consists of strongly antiferromagnetically coupled pairs of Cu^{2+} ions has received definitive proof in the case of ceruloplasmin. Thus *Van Leeuwen et al. (142)* have shown that when nitric oxide reacts with fully reduced ceruloplasmin an EPR signal is obtained which clearly arises from an $S = 1$ spin system and which exhibits a full compliment of $[2 \times 2 \times 3/2 + 1 =]$ seven hyperfine lines showing that two Cu^{2+} ions are in very close proximity in this form of the protein. By analogy with lacasses, the oxidized form of the diamagnetic Cu of ceruloplasmin may be confidently associated with the 330 nm absorption band first discovered by *Holmberg* and *Laurell (101)*.

Accessibility of Copper to External Perturbants

Since it is well known that at least part of the Cu of ceruloplasmin undergoes cyclic reduction by substrate and oxidation by molecular oxygen *(88)*, the availability of the various Cu binding sites to small solute molecules is of importance. One of the first observations which suggested that at least part of the Cu was

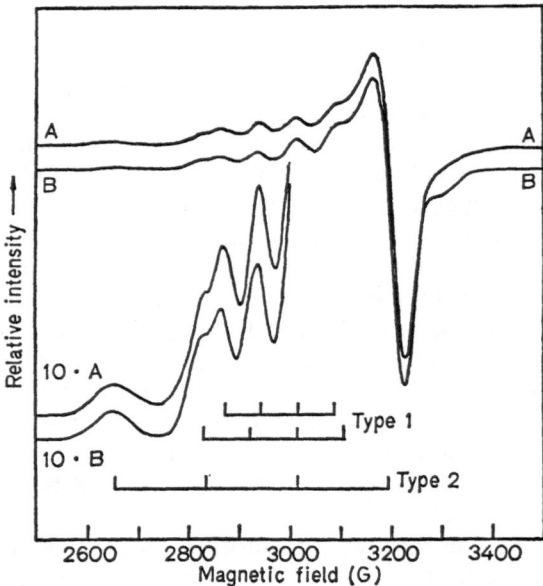

Fig. 6. Experimental (A) and computed (B) EPR spectra of human ceruloplasmin. The spectrum consisted of two Type 1 and one Type 2 ions. The simulated spectrum is the sum of three components using the following parameters. For the single Type 2 Cu^{2+} $g_{\parallel} = 2.247$, $g_{\perp} = 2.06$, $A_{\parallel} = 180$ gauss, $A_{\perp} = 25$ gauss; for one Type 1 ion, $g_{\parallel} = 2.215$, $g_{\perp} = 10$ gauss, $A_{\parallel} = 92$ gauss, $A_{\perp} = 10$ gauss; for the other Type 1 component $g_{\parallel} = 2.206$, $g_{\perp} = 2.05$, $A_{\parallel} = 72$ gauss, and $A_{\perp} = 10$ gauss. Gaussian linewidths (gauss) assumed were 70, 42, and 32 respectively. [Taken from *Gunnarsson, Nylen*, and *Pettersson*; Ref. (*141*)]

available to external ligands was the observation by *Holmberg* (*100*) that 1 mM KCN inhibited the oxidase activity toward paraphenylenediamine. The relationship between paramagnetic Cu^{2+} and solvent water molecules has been investigated by *Koenig* and *Brown* (*20*) who observed that protein bound Cu^{2+} had only a very small influence on the T_1 of solvent protons. From an analysis of relaxivity changes over the Larmor frequency range 0.01 to 50 MHz, it was concluded that a water molecule is bound between 3 and 7 Å from at least one Cu^{2+} center which in turn must be buried approximately 10 Å below the protein surface, and the residence time of the H_2O molecule at its binding site was estimated to be $\sim 10^{-5}$ sec. *Blumberg et al.* (*143*) in 1963 studied the relaxivity at 60 MHz as a function of the degree of bleaching of the blue color and found that plots of relaxivity vs. normalized absorption at 610 nm were sharply biphasic. These results, which are not readily rationalized in terms of simple contributions to the relaxivity from different yet independent Cu^{2+} ions of ceruloplasmin are widely quoted as indicating that the Cu^{2+} ions are physically different; they would further imply that the reduction of bound Cu^{2+} occurs in stages rather than in an all-or-none process.

Andréasson and *Vänngård* (*179*) have presented convincing evidence that azide ion binds to Type 2 Cu^{2+}, and they have suggested that this effects inhibition of oxidase activity. Results concerned with the relation between ion binding and inhibition will be reviewed below in conjunction with those dealing with the oxidase activity of ceruloplasmin.

A number of studies have dealt with the binding of anions to ceruloplasmin. *Kasper (144)* demonstrated that azide, cyanide, thiocyanate, and cyanate at fairly high concentrations produced spectral alterations of ceruloplasmin by liganding a portion of the Cu. He also found that approximately 12% of the total Cu could be removed irreversibly by 0.1 M KCN with only a 50% loss in oxidase activity. *Byers et al. (72)* have carried out an extensive study of the binding of several anions to the oxidized form of ceruloplasmin. Their results suggest the presence of two binding sites for azide and thiocyanate, with the stability constant of the second being approximately 100-fold less than the first. The authors have interpreted their data in terms of anation of the Type 3 Cu atoms for all anions except F^- which binds to the Type 2 Cu^{2+}.

By analogy with the behavior of Type 1 Cu^{2+} in the simpler blue proteins and considering the results of the proton relaxation studies it might be supposed that the Type 1 Cu^{2+} ions of ceruloplasmin would be unable to coordinate externally added ligands. However, *Wever et al. (145)* have reported that nitric oxide reversibly bleaches the blue color of the oxidized form of protein and they have proposed that nitric oxide combines with all Type 1 Cu^{2+} to form a diamagnetic Cu^+ --- NO^+ charge transfer complex. The remaining EPR signal has parameters similar to Type 2 Cu^{2+} and its intensity is 25—26% of the original signal. These experiments suggest that NO· may coordinate directly to Type 1 Cu since NO^+, necessarily formed upon reduction of the Cu^{2+}, is known to be a very reactive nitrating agent, and the reversibility or the above reaction suggests that it is probably not released into solution. However, the possibility that the two Type 1 Cu^{2+} are reduced by NO· via an intramolecular route cannot be excluded.

If all the assumptions which have been made in the interpretation of the above described perturbation experiments are correct, all Types of Cu are accessible to perturbants; this seems very unlikely.

Removal of Copper and Reconstitution

Holmberg and *Laurell (102)* demonstrated that copper could be removed from ceruloplasmin in an apparently irreversible manner. *Morrell* and *Scheinberg (195)* later demonstrated that when ascorbate reduced ceruloplasmin was treated with diethyldithiocarbamate, copper was removed with concomitant formation of a colloidal suspension of Cu^+-diethyldithiocarbamate which could be removed by high speed centrifugation leaving a clear but slightly yellow supernatant solution. Appropriate treatment of this solution with an ion exchange column yielded a water-clear solution containing a large portion of the original protein. Ceruloplasmin could be reconstituted in 95% yield by added Cu^{2+} and ascorbic acid. In this study only optical properties and the oxidase activity were used as measures of reconstitution. *Aisen* and *Morrell (196)* showed that reconstitution was an all-or-none process. Thus, if less than a full complement of Cu^{2+}/ascorbate was present only fully regenerated ceruloplasmin and apoprotein were found to be present. This can only occur if the binding of copper to apoprotein is cooperative. Indeed, there must be a rather strong stabilizing interaction between the several

Cu-binding sites, and this serves as one of the better examples of protein mediated Cu-Cu interaction in the multicopper oxidases. These authors also established that the EPR and immunochemical properties of reconstituted ceruloplasmin were identical with those of native oxidized protein.

Morrell et al. (*197*) have shown that reconstitution to oxidized protein can be effected anaerobically in the presence of Cu^{2+} and cysteine. Maximal blue color was developed, relative to an aerobic control, when sufficient cysteine was present to reduce approximately 63% of the Cu^{2+} initially present. Under these conditions, introduction of oxygen led to only a small increase in blue color. The authors did not indicate whether the added cysteine was in fact oxidized in these experiments, and it is possible that cysteine serves to facilitate the incorporation of Cu^{2+} rather than simply reducing it to Cu^+. These experiments showed that the blue color was not dependent on the presence of molecular oxygen.

4. Oxidation - Reduction Properties. *How many electron acceptors in ceruloplasmin*? Early results relevant to this question suggested that a number of reducing equivalent corresponding to half the total Cu present was adequate to completely decolorize ceruloplasmin solutions (*143, 146, 147*). Recently, however, conflicting results have been reported on this point, *Carrico et al.* (*120*) and *Deinum and Vänngård* (*123*) have published results showing that the capacity of the protein to bind electrons is 1/Cu. The rather high redox potentials (*123*) of these acceptors suggest an analogy with the laccases, and the electron accepting sites may be reasonably associated with Cu^{2+} ions. In contrast *Veldsema* and *Van Gelder* (*140*) report results indicating that four electrons can be bound by each molecule of ceruloplasmin. This would nominally correspond to 0.5 e⁻/Cu as these authors have assumed 8 Cu/molecule.

Some clues to the basis of these discrepancies are found in the conditions used by the different workers. *Carrico et al.* (*120*) carried out most titrations in acetate buffers of pH 6.0 having an ionic strength of 0.05. Under these conditions ceruloplasmin accepted 1 e⁻/Cu. In 0.1 M NaCl and 0.5 M sodium phosphate at pH 7.4 only 0.5 e⁻/Cu were accepted. [These conditions were apparently the same used by *P. Aisen* in unpublished experiments, cf. Personal Communication referred to in Ref. (*120*)]. Thus, it appears that high concentrations of phosphate affect the electron accepting capacity of the protein. However, the situation may be more complex as *Veldsema* and *Van Gelder* (*140*) titrated the enzyme at two pH values, 5.2 and 7.0, in the presence of 0.3 M sodium acetate finding 0.5 e⁻/Cu while *Deinum* and *Vänngård* (*123*) find 1 e⁻/Cu under identical conditions. *Veldsema* and *Van Gelder* (*140*) also report that treatment of their ceruloplasmin preparations with 5 mM p-chloromercuribenzene sulfonate results in a modification such that an electron binding capacity of 1/Cu is observed. Perhaps the point on which the dilemma hinges is the variable rate at which electrons are accepted by the molecule depending on its environment and history. The time required to bind one electron per Cu varies from 15 min to 10 hours (*120. 123*), While all the results cannot be clearly rationalized on the basis of the published observations and conditions, the variability of reactivity toward reducing agents in the presence of phosphate and acetate buffers the former of which is a known inhibitor of the oxidase activity (*148*) and the modification of the electron binding capacity by mercurials are quite similar to the observations of *Witwicki* and *Zakrzewski* (*126*) on the salt dependent reactivity of the single sulfhydryl group available for reaction in all but freshly prepared samples of ceruloplasmin (55). Taken together the results suggest that the ceruloplasmin molecule is conformationally sensitive to its specific ionic environment such that its reactivity toward both mercurials and substrates is drastically modified. It is also quite possible that its sensitivity to ionic perturbations is strongly dependent on the degree of proteolysis which has occurred. The above considerations may be extended to suggest that the ceruloplasmin molecule exists in at least two distinct interconvertible conformational isomers having different properties. For example, one form can accept one electron per Cu and the other can accept approximately half an electron per Cu atom.

Redox Potentials of the Electron Accepting Sites in Ceruloplasmin

The early observation of *Holmberg* and *Laurel* (101) that ferricyanide could not reoxidize ascorbate reduced ceruloplasmin was the first indication that the blue color center had a redox potential considerably greater than the ferri/ferrocyanide couple ($\sim$430 mV). This observation led *Morell et al.* (197) to determine that the blue color of ceruloplasmin did not require the presence of strongly bound molecular oxygen (cf. Section C3). Somewhat later it was shown that the redox potential of the blue color was between 500 and 600 mV (63). In an extensive study of this subject *Deinum* and *Vänngård* (123) have examined the spectral properties of ceruloplasmin at equilibrium potentials ranging from 450 to $\sim$610 mV. Based on an analysis of significant curvature in the Nernst plot, the authors concluded that there were two, nominally independent, Type 1 Cu^{2+} ions each having identical molar absorbance indexes at 610 nm but having redox potentials of 480 and 590 mV. Thus, regardless of the uniqueness of this interpretation ceruloplasmin can be considered to have two strongly oxidizing one electron centers.

Attempts to correlate changes in 330 nm absorbance with equilibrium redox potentials were uninformative (123). Nernst Plots yielded n-values ranging from 0.5—1 relative to the Type 1 centers. Indeed, *Deinum* and *Vänngård* (123) question whether the 330 nm band is a reliable indication of the redox state of the non-paramagnetic Cu atoms. This has also been pointed out by *Veldsema* and *Van Gelder* (140) who have suggested that absorbance changes at 330 nm, which can be maximal even when 0.5 e^-/Cu have been taken up by the protein (120), are a reflection of a conformational change in the region of the diamagnetic Cu atoms.

Since the 330 nm band is evident only in the oxidized minus reduced difference spectrum (101), and other chromophores, particularly disulfide bonds having a dihedral angle significantly different from 90° (70, 149, 150) may absorb in this region, it is reasonable to question whether Type 3 Cu^{2+} is responsible for all the absorption at 330 nm. It is possible that the absorbance changes observed in this region during electron titrations of the 1e^-/Cu conformation of ceruloplasmin are due to both the reduction of the EPR-nondectable $(Cu^{2+})_2$ pairs and to changes in the disulfide bonds, while electron titration of the 0.5 e/Cu form, in which only the paramagnetic Cu^{2+} ions are reduced, could be due to conformational changes which affect both chromophores.

5. Mechanism of the Oxidase Activity. Considerable work has been done toward elucidating the mechanism of action of ceruloplasmin's oxidase activity.

There have been basically two approaches to the problem: the use of classical steady state kinetic analyses and the application of the various methodologies applicable to the brief transient phase as the enzyme-substrate mixture approaches the steady state or independently interacts with either a reductant or oxygen.

Steady State Studies

Two major experimental difficulties cloud the interpretation of the kinetic results. The first and least important of these is the effect of intermediary free radicals as substrates of the enzyme, and the second and most important is the role of iron salts in both the steady state kinetics of ceruloplasmin mediated oxidations of various substrates and on the rates of the individual reactions between substrates and enzyme. The latter is in contrast to *Rhus* laccase whose kinetic behavior is uninfluenced by free iron (97).

Two of the most commonly used substrates of ceruloplasmin are paraphenylenediamine (PPD) and N, N-dimethyl-paraphenylenediamine (DPD). DPD, which has been used most extensively, is oxidized in a one-electron process to DPD+ which is intensely colored and allows ready observation of the course of the reaction. It has been well established that both substrates are initially oxidized to their respective radical cations. *Broman et al.* (88) studied the oxidation of PPD, observed free radical formation during catalysis, and identified this species as PPD+ by EPR techniques. Very convincing evidence was also presented that this free radical was neither bound to the enzyme nor acted as a substrate toward ceruloplasmin, but instead rapidly disproportionated to PPD and the quinoid form of PPD.

$$C_p + PPD \longrightarrow C_p^- + PPD^+$$
$$C_p + PPD^+ \xrightarrow{\;\;\;\times\;\;\;}$$
$$\text{non-enzymatic}$$
$$2\,PPD^+ \underset{\text{(rapid)}}{\rightleftharpoons} PPD + PPD^{++}$$

The same reaction scheme does not hold for DPD where the cation radical, DPD+, also serves as a substrate for ceruloplasmin.

$$C_p + DPD^+ \longrightarrow C_p^- + DPD^{++} \text{ (colorless)}.$$

Clearly, if the latter reaction occurred to a significant extent complications would arise in the interpretation of kinetic results. Indeed, Lineweaver-Burk plots were found to be non-linear where DPD was used as substrate (151—154), and initially this was interpreted as evidence for the presence of two active sites for reducing substrate. *Walaas et al.* (151) provided good evidence that DPD+ could act as an effective electron donor to ceruloplasmin, but they did not recognize that this could lead to non-linear Lineweaver-Burk plots and maintained that there were two different sites.

Pettersson and his co-workers (155—160) have derived a rate law for the ceruloplasmin catalyzed oxidation of DPD on the basis of a model which involved only a single active site on the enzyme, assumed that the reaction, $2\,DPD^+ \rightleftharpoons DPD$

+ DPD^{++} was always at equilibrium, and that DPD$^+$ could reduce ceruloplasmin; the derived rate law accurately fit the experimental data. The curved Lineweaver-Burk plots previously observed therefore resulted from the fact that DPD$^+$ was acting as a substrate, and therefore ceruloplasmin probably has only a single active site for substrate oxidation or possibly two independent and nominally identical sites.

The second major problem in the interpretation of the kinetics of ceruloplasmine catalyzed oxidations involves the unique role played by trace amounts of iron which contaminate the reaction mixtures. *Curzon* and *O'Reilly (161)* observed that both Fe^{2+} and to a lesser extent Fe^{3+} at concentrations stoichiometric with enzyme significantly enhanced the oxidase activity of ceruloplasmin toward DPD, and they suggested that a number of the conflicting results in the literature were due to variations in the concentration of contaminating iron salts. Iron appears to be unique in this regard since Mn^{2+}, Co^{2+}, Ni^{2+}, and Zn^{2+} activated to no more than 50% below 20 μM concentration while at higher concentrations the above divalent ions showed inhibition *(148, 162)*.

The observations of *Curzon* and *co-workers* on the effects of iron salts stimulated a great deal of work on the possible involvement of trace iron in a variety of ceruloplasmin oxidations. *Levine* and *Peisach (162)* demonstrated that iron is a common contaminant of DPD and that the activity of the enzyme toward this substrate was strongly inhibited by micromolar concentrations of EDTA. Indeed the inhibitory effect of EDTA on DPD oxidation decreases very rapidly near 10^{-6} M *(162)* as was first demonstrated by *Curzon (148)* while when psilocin (N, N-dimethyl-4-hydroxy tryptamine) was used as a substrate, maximal inhibition occurred above 10 mM EDTA. Indeed, in this system, EDTA does not appear to be acting as a chelating agent toward trace metals but more as an anionic inhibitor *(148)*. This provided some indication that the stimulatory effects of Fe^{2+}/Fe^{3+} may be largely dependent on the nature of the substrate. Further support of this contention is found in the results of *Levine* and *Peisach (162)* who showed that the concentration of Fe^{2+} yielding maximal activation was strongly dependent on the initial concentration of substrate and moderately dependent on the nature of the substrate. Also, it appears likely that some potential substrates, such as ascorbic acid, may only be oxidized if iron is present in the solution.

How much endogenous Fe salts are present in ceruloplasmin solutions and how effective are they as stimulators of ceruloplasmin oxidase activity? *McDermott et al. (163)* have very elegantly demonstrated that Chelex treated solutions of ceruloplasmin were between 1 and 40 nmolar Fe depending on the amount of Fe the solution contained prior to treatment with Chelex (2—20 Fe atoms per 10,000 molecules of protein). At these extremely low concentrations of iron and ratio of iron to ceruloplasmin it might be expected that Fe would play no significant role. However, it was shown that the ascorbate oxidizing activity of the protein nearly doubled when the concentration of Fe in the stock solution of ceruloplasmin increased from 2—14 nanomolar.

There has been some question whether ascorbate is in fact a substrate of ceruloplasmin. *Morell et al. (164)* concluded that ceruloplasmin itself was incapable of oxidizing ascorbic acid but that traces of Cu^{2+} were responsible for the apparent catalysis. *McDermott et al.* (*163*) have provided additional evidence that ceruloplasmin is not an ascorbate oxidase but suggest that trace Fe rather than Cu is responsible for apparent activity. *Gunnarsson et al.* (*166*) argue that ascorbate is in fact a substrate of the iron free enzyme system. Inasmuch as ascorbate can reduce all available electron centers in the enzyme (*120*) it seems unreasonable that the protein would not have some oxidase activity toward this substance.

McDermott et al. (*163*) have proposed that there are two groups of substrates for ceruloplasmin: those substances for which non-chelated Fe is not required and therefore which interact directly with the enzyme and those substrates which require the presence of non-chelated Fe to be catalytically oxidized. Some substrates belonging to the former group are epinephrine, PPD, norepinephrine, dopamine, and serotonin. Those falling in the second class are ascorbate, hydroquinone, ferrocyanide, L-DOPA, hydroxylamine, thioglycolic, acid and cysteine.

The present interpretation of these observations was formulated by *Curzon* and *O'Reilly* (*161*) in 1960 and elaborated by *Curzon* (*165*) and *McDermott et al.* (*163*). Thus

$$\text{Spontaneous} \longrightarrow \Big)\Big(\quad\Big)\Big(\longleftarrow \text{Ceruloplasmin.}$$

with SH_2, Fe^{3+}, H_2O above and $2H^+ + S$, Fe^{2+}, O_2 below.

However, as pointed out by *Gunnarsson et al.* (*166*) the kinetic limitations portended by a reaction between 10^{-8}—10^{-9}M "free" Fe^{3+} and substrate have not been adequately explored, and the suggestion of *Levine* and *Peisach* (*162*) that Fe-substrate complexes may be more readily oxidized by enzyme than the substrate alone should not be excluded [cf. Ref. (*227*)].

Cognizant of the above complications in the interpretation of the kinetics of ceruloplasmin mediated oxidations, *Young* and *Curzon* (*167*) have measured V_{max} and K_m for 37 different organic substrates. These workers used a reaction mixture containing substrate, excess ascorbic acid, 100 μM EDTA, and sulfate was maintained at 10 mM as a replacement for chloride which acts as an inhibitor; the progress of the reaction was followed with an oxygen electrode. Only when it was nearly depleted did oxygen become limiting, as *Frieden et al.* (*168*) have shown that the K_m of O_2 is approximately 4 μM. The function of the ascorbate is to rapidly reduce potentially long-lived cation radicals which themselves may act as substrate or inhibitors of activity. This procedure effectively couples ascorbate oxidation to that of substrate.

The data of *Young* and *Curzon* (*167*) are of prime importance in understanding the mechanism of the oxidase activity of ceruloplasmin. Of the 37 substrates tested V_{max} values varied over only an 8-fold range (1.3—10.8 ē/Cu-min at pH 5.5 in 10 mM sodium acetate; k_{cat} = 0.15—1.26 sec^{-1}) and for the 20 para-aminophenol compounds the total variance in V_{max} was only a factor of two. In contrast, K_m for these substrates varied over a range of 10^4 (20 μM—0.28 M). (Table 5.) The inescapable conclusion from these data is that the rate limiting step in ceruloplasmin oxidase activity is *independent of the nature of substrate*. Indeed, even the small variations in V_{max} might be attributed to minute quantities of

Table 5. Steady state kinetic parameters for substrates of ceruloplasmin[a]

Substrate	V_{max} (ē/Cu-min)	K_m (Micromolar)
p-Amino compounds		
N-Acetyl-p-phenylenediamine	3.42	12300
p-Aminophenol	3.53	1540
p-Anisidine	4.05	6140
2-Chloro-p-phenylenediamine	5.35	241
NN'-Di-s-butyl-p-phenylenediamine	6.05	620
2,5-Dichloro-p-phenylenediamine	3.94	740
NN-Diethyl-p-phenylenediamine	3.18	556
NN'-Dimethyl-p-phenylenediamine	4.33	164
NN-Dimethyl-p-phenylenediamine	5.13	203
Durenediamine	6.00	171
N-Ethyl-N-(2-hydroxyethyl)-p-phenyl-enediamine	5.49	110
N-Ethyl-N-2-(S-methylsulphonamido)-ethyl-p-phenylenediamine	6.06	87.2
2-Methoxy-p-phenylenediamine	6.20	161
N-(p-Methoxyphenyl)-p-phenylenediamine	6.50	20.6
2-Methyl-p-phenylenediamine	5.49	213
2-Nitro-p-phenylenediamine	6.93	1260
p-Phenylenediamine	4.44	292
p-Phenylenediamine-2-sulphonic acid	3.83	2660
N-Phenyl-p-phenylenediamine	4.78	47.7
NNN'N'-Tetramethyl-p-phenylenediamine	5.06	197
Catechols		
L-Adrenaline	2.29	2550
Catechol	8.98	282000
3,4-Dihydroxyphenylacetic acid		⩾ 250000
3,4-Dihydroxyphenylalanine (dopa)		∼ 20000
3,4-Dihydroxyphenethylamine	7.54	2850
4-Methylcatechol	6.81	60300
L-Noradrenaline	2.71	2810
5-Hydroxyindoles		
5-Hydroxyindol-3-ylacetic acid	1.50	8340
5-Hydroxytryptamine	5.68	908
5-Hydroxytryptophan	1.78	16300
5-Hydroxytryptophol	2.87	5100
Other compounds		
m-Aminophenol	4.03	199000
o-Aminophenol	3.60	2880
Ascorbate	4.07	5200
NN-Dimethyl-m-phenylenediamine	4.02	3050
NN-Dimethylaniline		⩾ 25000
p-Methoxyphenol		⩾ 700000
m-Phenylenediamine	5.60	36000
o-Phenylenediamine	1.30	2950
Pyrogallol	10.80	57900
Quinol	5.64	65700

Table 5 (continued)

Substrate	V_{max} (ē/Cu-min)	K_m (Micromolar)
Fe^{2+}[b])	22	0.6
Resorcinol (100 mM)		
Gentisic acid (100 mM)	Activity undetectable	
Aniline (2 mM)		
Mercaptoethanol (100 mM)		

[a]) All data taken from *Young* and *Curzon* (*167*) unless otherwise indicated. This reference should be consulted for details.

[b]) From *Osaki* (*169*).

uncomplexed Fe and other complications associated with the complex reaction mixtures.

Osaki (*169*) has studied the catalytic oxidation of ferrous ion by ceruloplasmin, and reported non-linear Lineweaver-Burk plots explicable in terms of two K_m values: 0.6 μM and ∼50 μM. This behavior could arise from the fact that ceruloplasmin may bind Fe(III) as it is known to bind several other metal ions quite strongly (*170*), and does not provide evidence for more than one type of active site. Ferrous ion is an excellent substrate of ceruloplasmin as evidenced by its very low K_m. However, its V_{max} does not differ greatly from the values of all the substrates tested by *Young* and *Curzon* (*167*). From the data of *Osaki* (*169*) V_{max} at pH 5.5 is 30 ē/Cu-min at 30 °C (k_{cat} = 3.5 sec^{-1}). Using the activation energy of 10.9 Kcal/mole reported by *Osaki* and *Walaas* (*171*) for the initial velocity of Fe(II) oxidation the above V_{max} can be corrected to 22 ē/Cu-min at 25 °C. Thus, it appear that for all substrates of ceruloplasmin V_{max} is remarkably independent of the nature of the substrate.

Transient Kinetic Studies

The early suggestion of *Holmberg* and *Laurell* (*102*) that Cu was involved in the oxidase activity of ceruloplasmin, possibly undergoing cyclic reduction by substrate and oxidation by molecular oxygen, has received ample verification by a number of investigators (*88, 172, 173*). Kinetic studies of the reduction of ceruloplasmin and reoxidation by molecular oxygen have revealed several qualitative features of the reaction: (a) The rate of reduction of Type 1 Cu^{2+}, even by so-called iron mediated substrates such as ascorbic acid, and spectral changes which occur in the 330 nm region occur very much faster than the rate limiting step in the overall oxidase activity. (b) In the presence of a slight excess of reducing equivalent (∼8 ē/molecule) the rate of oxidation of approximately half the Type 1 Cu^+ and the spectral changes in the 330 nm region also occur very rapidly. (c) During reduction no spectrally distinct intermediates are apparent, but on reoxidation an intermediate forms having an absorption maximum at 420 nm. (d) The rate of decay of this intermediate occurs at a velocity comparable to the turnover rate of the enzyme system, under substrate saturating conditions.

41

Studies on the rate of reduction of ceruloplasmin have involved some substances which according to *McDermott et al.* (*163*) require iron salts to act as substrates, and it appears that iron may also play a role in the transient approach, to the steady state as well as the overall rate under conditions of both O_2 and substrate saturation. *Osaki* (*169*) has observed that iron chelating agents such as 1,10 phenanthroline and apotransferrin quite dramatically impede the reduction of Type 1 Cu^{2+} by ascorbate. For example, a 16.5 μM, Chelex treated, ceruloplasmin solution was half-reduced in a $\sim$3 sec time period in the absence of apotransferrin and $\sim$7 sec in the presence of 20 μM apotransferrin, while 1,10-phenanthroline at 0.3 mM extended this time to $\sim$13 sec. It is possible that phenanthroline may inhibit by combining with the enzyme, but this seems very unlikely for apotransferrin which probably inhibits by sequestering the small amount of remaining Fe^{3+}. These observations suggest that results dealing with the rate of reduction of ceruloplasmin will not be quantitatively reliable unless endogenous iron is adequately bound by a chelation system. *Manabe et al.* (*173*) state that EDTA also significantly retards the rate of reduction of ceruloplasmin by ascorbate.

Gunnarsson et al. (*166*) have measured the rate of reduction of Types 1 and 3 Cu^{2+} in the presence of excess EDTA by several substrates, including ascorbate. Both reactions were found to be well behaved second-order processes in which no saturation was observed even when substrate was present at a concentration several-fold greater than its steady state K_m value. The rate of reduction of Types 1 and 3 were identical suggesting either a complete equilibration between the two electron acceptors or identical rates of reduction by substrate. When less than stoichiometric amounts of substrate (ascorbate) were present, however, the two chromophores were reduced at significantly different rates (*172*).

These authors demonstrated that a very simple relationship exists between V_{max} and K_m obtained from steady state experiments and k_1 obtained from transient studies.

Assuming that the simple kinetic scheme holds

$$\text{E}_{ox} + \text{S} \xrightarrow{k_1} \text{E}_{red} + \text{P}$$
$$\text{E}_{red} + \text{O}_2 \xrightarrow{k_2} \text{E}_{ox}$$

$V_{max} = k_2$, $K_m = k_2/k_1$, and $V_{max}/K_m = k_1$. Observed and calculated values of k_1 are given in Table 6. It is evident that for the more slowly reacting substrates $V_{max}/K_m \equiv k_1$ obtained from transient studies. The authors point out that the apparent lack of correspondence with the faster reacting substrates is probably due to difficulties in the direct measurement of k_1. As might be expected, there is an excellent correlation between k_1 (obsd) and K_m^{-1}, supporting the relatively simple assumptions regarding the rate limiting processes. This correlation may be explained in terms of an enzymatic reaction which does not involve formation of the classical E · S complex, but only an encounter complex of E and S in which they exchange an electron.

Levine and *Peisach* (*174*) have reported a correlation of activity toward a particular substrate and the sum over its substituent Hammett σ-constants, indicating that an increased electron density in the aromatic ring results in

Table 6. Relation between transient and steady state parameters for various substrates of ceruloplasmin (166)[a]

Substrate	k_1 M^{-1} sec^{-1} Transient	$\dfrac{V_{max}}{K_m}$ $(= k_1)$[b] Steady state
Catechol	2.9 ± 0.4	3.2
Quinol	7.6 ± 1.2	8.6
Ascorbate	70 ± 10	80
Epinephrine	100 ± 10	90
p-Aminophenol	120 ± 15	230
PPD	1050 ± 100	1500
DPD	1600 ± 100	2500

[a]) 25 °C, 0.05 M acetate, 0.5 mM EDTA, and 0.25 mM O_2.
[b]) K_m and V_{max} taken from Ref. (167) (Table 5).

greater oxidizability. However, as pointed out by *Curzon* and *Speyer* (175) the Hammett σ relates the effect of one substituent at another carbon atom of the ring and does not indicate electron density on the ring. *Pettersson* (176, 177) and his co-workers have established an exponential dependence of K_m on the energy of the highest occupied molecular orbital of the substrate, which correlates with the ionization potential of the substrate (178) and a zero correlation of this energy with V_{max}. The results shown in Table 6 support the contention that the substrate specificity is primarily due to the contribution of k_1 to K_m. It is interesting that the predicted k_1 for Fe^{2+}, based on V_{max} and K_m values of *Osaki* (169) is $\sim 4 \times 10^6$ M^{-1} sec^{-1} and the observed constant (see below is $\sim 1 \times 10^6$ M^{-1} sec^{-1}.

The reoxidation of reduced enzyme by oxygen has been more extensively studied than its reduction by substrates. *Carrico et al.* (172) have reported that appearance of 610 nm absorbance occurs in two distinct phases; approximately 55% of total 610 nm absorbance reappearing 20 msec after mixing with oxygen and the remainder being completely developed only over a period of several hours. Absorbance changes at 340 nm followed a much more complicated path; after 15 msec it was several percent higher than the resting value, this was followed by a decrease for a period of approximately two seconds, and a subsequent slow increase then occurred to obtain the final expected value in about 5 min. The authors state that the excess absorbance appearing in the 300—450 nm region was indistinguishable from the spectrum of the 340 nm chromophore in the resting enzyme. *Manabe et al.* (173, 179) have appreciably extended the work of *Carrico et al.* (172), and have confirmed that reoxidation of Type 1 Cu occurs in two phases, one rapid and one much slower, and they have shown that the rate of the first phase depends on oxygen concentration while the slow phase does not, Fig. 7. During the reoxidation of reduced ceruloplasmin there is a transient appearance of an intermediate having an absorption maximum at 420 nm (179). The rate of appearance of this intermediate was linearly but not propor-

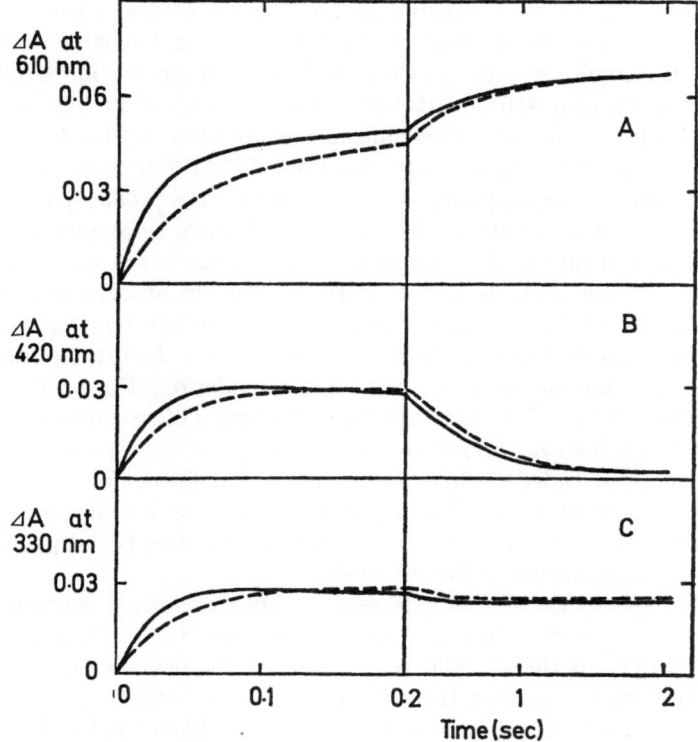

Fig. 7. Time course for the aerobic oxidation of reduced ceruloplasmin by molecular oxygen. The reaction conditions were pH 5.5, 0.05 M acetate, 50 μM EDTA, and 25 °C. The ceruloplasmin concentration was 14.7 μM, and the oxygen concentrations were 80 μM (dashed) and 530 μM (solid). [Taken from *Manabe, Hatano*, and *Hiromi*; Ref. (*173*)]

tionally related to oxygen concentration as were the concomitant changes at 610 and 330 nm, and the apparent second order rate constants measured at each wave-length were essentially identical over an approximate 10-fold range in oxygen concentration (*173*), consistent with the low K_m for O_2 ($\sim$4 μM). During the second phase of the reaction in which 610 nm absorbance increases and 420 nm absorbance decreases the rate constants obtained at both wave-lengths are identical and independent of oxygen. As will be discussed below, this reaction is thought to correspond to the rate limiting step of the overall reaction. The observation of transient absorbance at 420 nm contrasts with the report of *Carrico et al.* (*172*), and it was suggested (*173*) that the failure of these authors to find the 420 nm band may be due to the fact that a system to chelate iron was not used. The failure of *Carrico et al.* (*172*) to observe the 420 nm chromophore might indeed be explained by the recent observation of *Osaki* (*180*) that Fe^{2+} stimulates the slow phase of the reoxidation of Type 1 Cu^+.

As indicated above, Fe^{2+} is itself an excellent substrate of ceruloplasmin. *Osaki* and *Walaas* (*181*) have studied the reaction between Fe^{2+} and ceruloplasmin and its reoxidation by molecular oxygen after reduction by ascorbic acid. Their results may be summarized briefly: (a) Fe^{2+} reduced Type 1 Cu^{2+} in a second order process with a rate constant of $\sim 1 \times 10^6$ M^{-1} sec^{-1} at 20 °C, 0.2 M acetate pH 5.9. b) Concomitant with reduction of Type 1 an absorption band appeared with a maximum near 450 nm. However, *Carrico et al.* (172) were unable to reproduce this observation and its significance is not clear. (c) Pre-reduced enzyme was rapidly oxidized by O_2 with a constant of 0.57×10^6 M^{-1} sec^{-1}. This reaction was not reported to be biphasic, and the authors were probably only observing the rapid phase of reoxidation. (d) When 4 Fe^{2+}/mole ceruloplasmin are mixed in the presence of an excess of oxygen there is a rapid but not complete drop followed by a slow unimolecular increase in 610 nm absorbance. The authors have associated this slow return in color with the rate limiting step of the overall oxidase activity, and they have further suggested that the rate limiting step is a slow conformational change preceding rapid reoxidation. The results of *Manabe et al.* (173) support the viewpoint that the slow portion of reoxidation, not observed by *Osaki* and *Walaas* (181) in pre-reduced enzyme corresponds to the overall rate limiting step. Indeed, while the transient kinetic results do not mandate this step as rate limiting, the idea that this step involves a conformational alteration of the enzyme is consistent with and explains the fact that V_{max} is generally independent of the nature of the substrate.

The pH dependence of ceruloplasmin's oxidase activity was noted by *Curzon* (148) but was not studied in any detail. *Gunnarsson et al.* (177) have taken into account a variety of the complicating factors of the enzyme-DPD system, and they have obtained V_{max} over the pH range 4.5—6.5. Activity shows a maximum at pH 5.0 and gradually decreases toward zero at higher pH values. It would appear that an ionizable group on the enzyme, having a pK_a of ~ 6, is necessary for activity. At rather higher pH values, greater than 10, the physical properties of the enzyme are modified as a result of denaturation (198).

6. Inhibition of the Oxidase Activity. The oxidase activity of ceruloplasmin has been reported to be inhibited by a wide variety of different substances, and it would be well to briefly point out several ways in which an inhibition of the activity may be manifested: (a) A true inhibition of the activity resulting from a combination of enzyme or an enzyme-substrate complex with the effector. (b) Inhibition resulting from the sequestering of free iron, as has already been discussed. (c) Alteration of the chromophoric properties of the substrate or product by the effector, for example, the reduction of the colored radical cation of DPD. (d) Apparent inhibition will also result when the inhibitor causes aggregation or precipitation of the protein. Inhibition attributable to each of the above has been reported. [cf. *Curzon* and *Speyer* (175) for a lucid discussion of these points.] We will be concerned here primarily with three types of inhibitors, all of which act by a direct combination with some form of the enzyme: halide and inorganic anions, carboxylic acids having a double bond α- to the carboxyl function, and certain divalent cations.

Anions

Holmberg and *Laurell* (102) demonstrated that the oxidase activity was markedly affected by anions and these effects have been extensively studied by several investigators. It seems that all anions exhibit a degree of inhibitory action toward ceruloplasmin including anions of common proton buffer systems such as acetate and phosphate (148). The inhibition by both Cl^- and OAc^- was found to be of the uncompetitive class and OAc^- protected against the inhibition by Cl^- suggesting a rather complicated interaction between anions and the enzyme. Since acetate overcomes a significant portion of chloride inhibition, Curzon (148) suggested that is would be advantageous to use a fairly concentrated acetate buffer. While this offers a lower sensitivity it reduced interference by anions, such as Cl^-, which are frequently added stoichiometrically with substrate.

In 10 mM acetate buffer at pH 5.5 and with 4 μM EDTA present to reduce the catalytic effects of iron salts, the order of inhibition was determined to be: $CN^- > N_3^- > F^- > I^- > NO_3^- > Cl^- > Br^- > OCN^- > SCN^- > SeCN^- >$ (ClO_4^-, tetraborate, boric acid, phosphate, sulfate and cacodylate). Clearly the strongest inhibitors are those with metal binding capabilities although this features alone does not readily correlate the series. Indeed, there are complexities which are · apparent under detailed scrutiny and which suggest a very complicated pattern of inhibitory action by these anions. Curzon and his co-workers and other investigators have carried out detailed studies of the inhibition by azide, cyanide, the halides, and mixtures of various anionic inhibitors.

Azide

The steady state inhibition by azide is reversible and may be overcome by excess acetate, EDTA, or chloride (182). Lineweaver-Burk plots at various azide concentrations were linear and parallel which suggests that azide inhibits by binding to an enzyme substrate complex or to an intermediate form of the enzyme during catalysis (182). Gunnarsson et al. (183) have shown that azide, as well as other anions, does not inhibit by binding to an ES complex and therefore must exert its effect by combining with an intermediate form produced during catalysis.

The binding of azide to the inhibited form is quite strong, $Ki = 1.5$ μM, and by studying mutual depletion effects Curzon (182) was able to demonstrate that a single azide molecule was bound to inhibited ceruloplasmin. Inhibition by azide was found to be time dependent, becoming evident 30—60 sec. after addition of the substrate, and the lag period is substantially increased at lower temperatures. Pre-incubation of either the oxidized or reduced form of the enzyme with azide had no affect on this lag period (182, 183). The binding of azide to ceruloplasmin and the mechanism of its inhibition has been the subject of considerable discussion. Binding constants (expressed as dissociation constants) for the interaction between N_3^- and ceruloplasmin are given in Table 7. These constants were measured by a variety of techniques: appearance of an absorption band at 380 nm, alteration of that part of the EPR spectrum corresponding to Type 2 Cu^{2+}, and by inhibition of steady state activity. The wide range of values suggests the possibility that the different techniques are actually measuring different phenomena.

Andreasson and Vänngård (139) have correlated optical changes, EPR changes, and inhibition of steady state activity which occur upon interaction of ceruloplasmin with azide, and they have proposed that azide inhibits ceruloplasmin

Table 7. Comparison of dissociation constants for the binding of azide to ceruloplasmin

K_1 (Micromolar)	Type of measurement(s)	Experimental conditions	Ref.
67	Optical, EPR, Steady state kinetics	25 °C, 0.1 M acetate, pH 5.5, 2 mM Cl⁻	(139)
260	Binding kinetics, 380 nm changes	⎫ 25 °C, 0.05 M acetate	(184)
370	Equilibrium of 380 nm absorption	⎬	
3	Steady state kinetics	⎭	
1–2	Steady state kinetics	7.5 °C, 0.01 M acetate, 0.66 mM Cl⁻	(182)
2	Steady state kinetics		(183)

activity by binding directly to Type 2 Cu^{2+}. However, *Manabe et al.* (184) have found a very substantial difference between the inhibition constant and the binding constant of azide to oxidized protein. *Manabe et al.* (184), *Curzon* (182), and *Gunnarsson et al.* (183) find the inhibition constant to be near 2 μM, while *Andréasson* and *Vänngård* (139) report a value of 67 μM. The source of this discrepancy would not appear to be with the differences in experimental conditions but could be related to the history of the different ceruloplasmin preparations used and to different enzyme concentrations.

Manabe et al. (173, 179, 184) have examined azide inhibition by stopped-flow spectrophotometry, and they have demonstrated that azide does not inhibit the reduction of Type 1 Cu^{2+} by substrate. Azide weakly inhibits the rapid changes at 610 and 330 nm, and the appearance of the intermediate absorbing at 420 nm which occur on reoxidation by oxygen, but the disappearance of the 420 nm intermediate and the slow phase of Type 1 oxidation are concomitantly and strongly retarded in the presence of this ion, Fig. 8. Therefore, azide inhibits primarily by impeding the decomposition of an intermediate in the overall reaction as required by the steady state behavior of the enzyme systems (182), but this appears to be a reduced form of the enzyme not an E · S complex. Perhaps the inhibition is brought about by the binding of azide to a Type 2 Cu^{2+} in this species but verification of this point has not been forthcoming.

Cyanide

Reversible inhibition by cyanide is only slowly expressed in a mixture of enzyme and substrate. While azide inhibition required approximately two minutes for full expression, cyanide may require as much as 15 minutes (185). As found with azide, preincubation of cyanide with the oxidized or reduced forms of the enzyme does not affect the lag time of inhibition. This suggests that cyanide does not inhibit by binding to either the fully oxidized or fully reduced protein. In contrast to azide, cyanide shows nominally competitive inhibition with substrate for an intermediate form of the enzyme. Since cyanide, like azide, binds very strongly to ceruloplasmin the use of mutual depletion kinetics has indicated that two cyanide molecules bind to ceruloplasmin in the fully inhibited form.

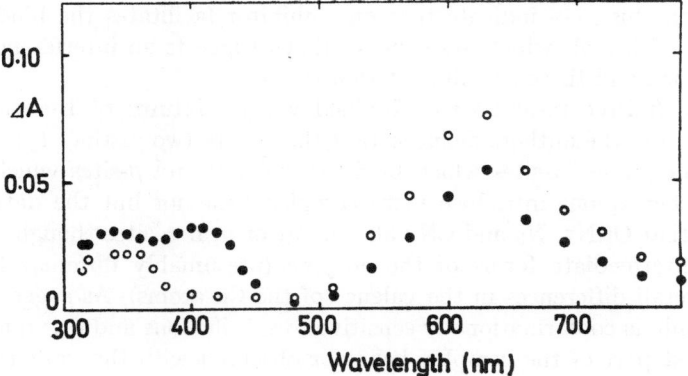

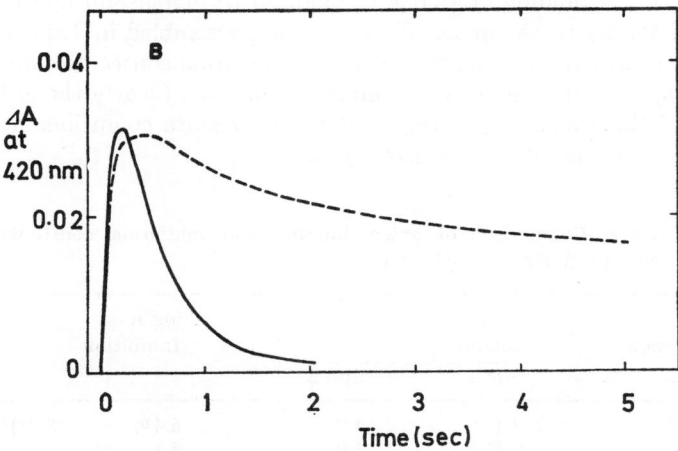

Fig. 8. Spectral changes observed in the reoxidation of reduced ceruloplasmin by molecular oxygen. (Upper) Spectra obtained at 0.2 (●) and 2 sec. (0) after mixing with oxygen. Note the absorption band in the region of 420 nm. (Lower) The affect of azide on the decomposition of the 420 nm band observed during reoxidation .The concentration of ceruloplasmin was 14.7 μM and azide was 120 μM. Conditions were the same as described in Fig. 7. [Taken from *Manabe, Manabe, Hiromi* and *Hatano*; Ref. (*179*)]

Inhibition by Mixtures of Anionic Inhibitors

The complexity of the inhibitory mechanisms of various anions was well demonstrated by *Curzon* and *Speyer* (*186*) who considered the inhibition of ceruloplasmin's oxidase activity by mixtures of various pairs of anions. Their data were presented in the form of inhibition isobols. Each point on an isobol representing, for example, 50% inhibition, is fixed by the concentrations of two anions at which 50% inhibition occurs. A straight line isobol indicates that both anions inhibit by binding to mutually exclusive sites. A curved, concave up, isobol indicates the inhibitors are binding to the same intermediate at different sites or, depending on the degree of inhibition this may indicate that one inhibitor facilitates the binding of the other. Curved isobols which are concave down suggests an interference between the inhibitors, and these are difficult to interpret.

Quite definitive results were obtained when mixtures of the halides were considered, and the authors deduced that there were two distinct types of halide anion binding sites: α-sites which bind Cl⁻ and F⁻; and β-sites which bind Br⁻ and I⁻. Other anions introduce more complex behavior but the data obtained suggested that OCN⁻, N_3^- and CN⁻ all bind at or near α-sites though possibly to different intermediate forms of the enzyme (presumably different forms may involve overall differences in the valence of the Cu atoms). As suggested above, ceruloplasmin is conformationally sensitive to specific ions and it is quite possible that at least part of the complex behavior observed with the various inhibitors may be due to a shifting of the enzyme among two or more conformational isomers.

The work of *Byers et al.* (72) and *Gunnarsson et al.* (183) allows quantitative comparison of the binding of anions to oxidized ceruloplasmin and inhibition of oxidase activity by these anions. These data are assembled in Table 8, and they show that no correlation exists between the association constant of an anion with oxidized enzyme and its ability to inhibit the enzyme. Clearly the anion binding properties of the enzyme operating under steady state conditions are different than when it is in the resting oxidized form.

Table 8. Comparison of anion binding and inhibition constants (association) for ceruloplasmin

Anion	log K Binding[a]		log K Inhibition	
	Site 1	Site 2		
N_3^-	3.4	1.7	5.4[b]	5.7[c]
SCN⁻	2.7	1.0	3.3	
OCN⁻	1.6	—	3.9	3.8
CN⁻	1.3	—	5.8	
Cl⁻	1.52[c]		1.8	2.0
F⁻	1.7[b], 1.6[c]		2.2	2.7
Br⁻	1.6[c]			1.8

[a] Taken from Ref. (72).
[b] Calculated by *Byers et al.* (72).
[c] Deduced from kinetic measurements (183).

Unsaturated Carboxylic Acids

It has been known for some time that certain carboxylic acids inhibit ceruloplasmin (109, 148). *Curzon* and *Speyer* (175) examined the inhibitory capacity of a number of types of organic substances including: aromatic carboxylates, unsaturated aliphatic carboxylates, aromatic non-carboxylates, unsaturated

J. A. Fee

aliphatic non-carboxylates, and a group of mono-substituted benzoates. All substances tested at 10mM concentration showed some degree of inhibition, ranging from $< 5\%$ to 70% for benzoate which was observed to be the best inhibitor. The functional groups which appear to enhance inhibition of a carboxyl group are an adjacent double bond with small substituents or an aromatic ring with no substituents. For example, ortho-methyl benzoate is a very weak inhibitor compared to benzoate. It was suggested that the carboxylic acid, like the inorganic anions, inhibited by binding to the metal ions. *Imoto et al.* (187) using NMR methods have demonstrated, however, that salicylate binds to both the fully oxidized and reduced forms of the enzyme but does not bind near Cu^{2+}. These authors refer to results which show that salicylate inhibits by decreasing the rate of reduction of enzyme by substrate. This is consistent with the studies of *Gunnarsson* and *Pettersson* (188) who concluded that carboxylic acids do not form $E \cdot I \cdot S$ complexes during inhibition but compete with substrate for the active site.

Metal Ion Inhibition

It has been shown that a variety of metal ions will inhibit oxidase activity at sufficiently high concentrations (148, 162). *McKee* and *Frieden* (170) have demonstrated that in addition to the Cu normally associated with native ceruloplasmin an additional 10 Cu^{2+}, 3 Fe^{2+}, 3 Zn^{2+}, 7 Ni^{2+}, or 16 Co^{2+} may be accommodated by the protein. The binding constants range from 2×10^5 to 2×10^4 M^{-1} and in the case of Ni^{2+}, Zn^{2+}, and Co^{2+} the inhibition constants of Fe^{2+} oxidation compare favorably with the dissociation constants of the metal-protein complexes. The inhibition mechanism is not understood.

Deuterium Oxide Inhibition

Osaki and *Walaas* (171) have observed that the oxidase activity of ceruloplasmin was reduced 42—43% in 66% deuterium oxide when Fe^{2+} was used as substrate. The authors suggested that proton-transfer may be involved in the rate limiting step, but since the effect of D_2O on real acidity was not evaluated it cannot be excluded that the inhibition merely reflects a change in acidity; this is an interesting observation.

7. **Summary.** The mechanism of the oxidase activity of ceruloplasmin is clearly very complex. Nevertheless, some overall features of the reaction have been firmly established: (a) Types 1 and 3 Cu are involved in an oxidation-reduction cycle, and it is possible that other Cu ions are likewise involved. (b) The catalytic rate under carefully defined experimental conditions, is quite independent of the nature of the substrate. Therefore, the limiting step must involve a transformation which occurs subsequent to reduction by substrate and presumably involves molecular oxygen. (c) The form of the enzyme that can react with O_2 does so with a very high affinity ($K_{m O_2} \sim 4$ μM). (d) The reaction between O_2 and fully reduced enzyme results in the formation of the complex composed of the elements of oxygen and enzyme and which absorbs 420 nm radiation. It appears that the rate limiting step is associated with the breakdown of this intermediate in the direction of products. The chemical nature of the intermediate is not understood

50

but it is reasonable to assume its structure is similar to that observed in the analogous laccase system (96).

The inhibition of ceruloplasmin is also quite complicated, and a part of this may be associated with the suggested conformational sensitivity of the molecule. Azide, at least, inhibits by decreasing the rate of decomposition of the 420 nm absorbing intermediate, possibly by inhibiting intramolecular electron transfer reactions necessary for the reduction of O_2 to H_2O.

D. Ascorbate Oxidase

The detailed chemistry and biochemistry of abscorbate oxidase have been adequately reviewed by *Dawson* and his co-workers (189, 190), and only a brief overview and a discussion of some recent developments will be presented here.

The reaction catalyzed by the enzyme is

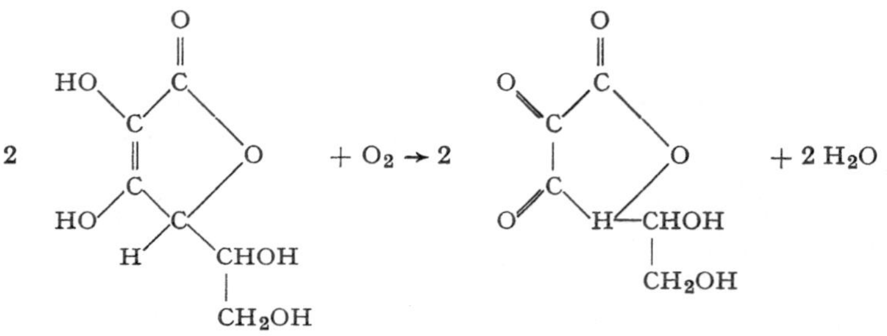

for which it shows a strong but not absolute specificity (190). *Yamazaki* and *Piette* (194) have demonstrated that ascorbic acid is oxidized via a one-electron process to yield the ascorbate radical. This behavior is analogous to the multicopper oxidases considered previously.

Largely due to the great difficulty associated with obtaining quantities of purified enzyme adequate for physical-chemical studies its characterization has not been carried to the same stage as that of ceruloplasmin and the laccases. Nevertheless, recent developments in the purification procedure (191, 192) have allowed measurements which signify a great similarity of the Cu-binding sites on this enzyme with those of other blue, multicopper oxidases. The properties of the enzyme are summarized in Table 2.

Recently *Strothkamp* and *Dawson* (54) have examined the quarternary structure of Zucchini squash (*Cucurbita pepo medullosa*) ascorbate oxidase by various denaturing procedures. Their findings suggest that the protein, having a molecular weight of approximately 140,000, consists of two readily dissociated subunits each of which consists of two peptide chains, A and B, which are covalently linked by disulfide bridges. The weight of the A chain was found to be 38,000 daltons and the B chain was 28,000 daltons. Barring any proteolytic digestion of the protein during preparation, these results suggest major differences between the overall structural properties of ascorbic acid oxidase and ceruloplasmin.

In spite of these potential differences in protein structure, the Cu atoms seem to be bound in environments substantially similar to those of the laccases and ceruloplasmin. Indeed, *Lee* and *Dawson* (*193*) and *Strothkamp* and *Dawson* (*54*) have deduced from a comparison of the optical properties of ascorbate oxidase with those of laccases that the former contains two Type 1 Cu^{2+} ions and four Type 3 Cu atoms. These account for six of the 8—10 Cu atoms usually found per molecule of protein and the remainder can be considered to be present in unknown proportions of Type 2 and extraneous Cu. The EPR spectra published by *Lee* and *Dawson* (*193*) and by *Nakamura et al.* (*191*) tend to support this conclusion. On the basis of these observations it has been suggested that ascorbate oxidase consists of two "laccase-type" units, but a deduction of this sort cannot be conclusive on the basis of present evidence.

Deinum et al. (*225*) have shown that the EPR spectrum of ascorbate oxidase from green zuchini squash is very similar to that from cucumber peel shown in Fig. 9. They have carefully recorded and integrated the EPR spectrum of ascorbate oxidase finding that $47 \pm 3\%$ of the Cu present was detectable by this method. Of this 47% only $25 \pm 5\%$ corresponded to Type 2 Cu^{2+}. With this information in hand, the authors quite accurately simulated the observed EPR spectrum assuming the presence of three EPR identical Type 1 Cu^{2+} and one Type 2 Cu^{2+}. On the basis of total Cu present and the above observations *Deinum et al.* (*225*) conclude that ascorbate oxidase contains three Type 1 Cu^{2+}, one Type 2 and four

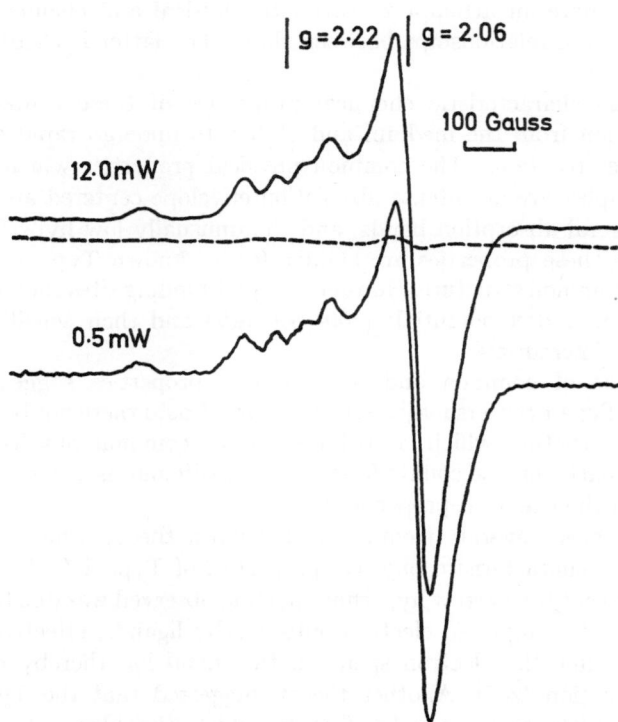

Fig. 9. The EPR spectrum of ascorbate oxidase from Cucumber. Native form (solid) and reduced with ascorbate (dashed). [Taken from *Nakamura, Makino* and *Ogura*; Ref. (*191*)]

Type 3 Cu. These conclusions are not consistent with the idea that ascorbate oxidase consists of two halves corresponding to laccase molecules.

Nakamura et al. (191) have carried out a fairly comprehensive study of ascorbate oxidase from cucumber (*Cucumis sativus*) peel. The Cu^{2+} was shown to undergo cyclic reduction and oxidation and the pH-activity curve was demonstrated to be bell-shaped with a broad maximum between pH 5.5 and 7. The velocity of the reaction depended on both ascorbate and O_2 concentrations with $K_m O_2 \sim 1$ mM and $K_m^{Asc} = 1.4$ mM at 25 °C. Azide was found to inhibit uncompetitively and $K_i = 0.2$ mM. Ascorbate oxidase appears to be substantially similar in its structure-function relationships to all the other known multicopper oxidases.

IV. General Comments on the Three Types of Copper Binding Sites

From the above discussion it is clearly possible to delineate the general properties of the various Cu centers, both physical and chemical. It is, however, except in the few instances difficult to rationally associate any given physical property to some structural feature which can be thought responsible for a particular chemical behavior. For example, while the intense blue color of Type 1 Cu^{2+} indicates an unusual structure, this and the other physical studies have not revealed a structural framework in which its chemistry can be rationalized. What follows then is more an attempt to correlate physical and chemical properties rather than propose relationships between these. The latter is clearly a task for the future.

Type 1. The characteristic chemical properties of these centers are their apparent isolation from the medium and ability to undergo rapid outer sphere electron transfer reactions. The common physical properties which signify this type of Cu complex are an intense absorption envelope centered around 600 nm made up of several absorption bands, and the unusually low hyperfine coupling constant, A_{11}. These properties are similar for all known Type 1 centers, and they suggest a common structure. However, Type 1 binding sites show considerable variation in their redox potentials (200—800 mV) and their sensitivity toward denaturation by mercurials.

The presence of common and non-common properties suggests that the structure of all Type 1 centers may not be the same. Could there not be an invariant portion of the structure which contributes to the common physical properties and a variable part which accounts for the minor differences among the physical parameters and the chemical properties?

There have been several attempts to develop a theory which satisfactorily accounts for the characteristic physical properties of Type 1 Cu^{2+}: The first of these suggested that the narrow hyperfine splitting observed was due to an unusual delocalization of the unpaired electron onto nearby ligands, effectively reducing the fraction of time the electron spent on the metal ion thereby reducing the hyperfine interaction *(201)*. Another theory suggested that the Type 1 center consisted of a Cu^{2+} and a nearby Cu^+ in which the blue color arose from a Cu^+ to Cu^{2+} charge transfer transition and the narrow hyperfine splitting from a form of delocalization *(143)*. This theory was discarded with the discovery that proteins containing only one Cu^{2+} were intensely blue and exhibited the characteristic EPR spectrum (202, 88). Other theories (203, 204) have been proposed which accounted for some of the physical properties, but absorption in the region of 600 nm was thought to arise from d—d transitions which were allowed to an unusual extent. The recent observations indicating the presence of more absorption bands than can be accounted for as d—d transitions complicates these theoretical interpretations. It is fair to say that an adequate theory does not at present exist. *Vänngård* (205) has recently evaluated the various theories and the lack of success in formulating model complexes which emulate the properties of the Type 1 center.

It is reasonable to ask what the chemical studies can tell us about the nature of the Type 1 center. There have been repeated indications that a unique feature of this center was a Cu^{2+} liganded to cysteine sulfur. Thus, the original observation of *Katoh* and *Yakamiya* (37) that spinach plastocyanin was decolorized by mercurials was also found to be true for a number of other simple blue proteins. However, ceruloplasmin and laccases and ascorbate oxidase were found not to be denatured by mercurials. In the case of *Rhus* laccase it was elegantly demonstrated that its single SH group was independent of Type 1 Cu^{2+}. These observations leave us with the dilemma that either Type 1 centers are not intrinsically the same or if they are the cysteine sulfur does not serve as a direct ligand to the copper. It is, of course, possible to rationalize the sensitivity to mercurials with a structure having an SH group in close proximity to but not directly coordinated to the copper.

The following amino acids have been definitely excluded as part of a common structure of the Type 1 center: Tryptophan has been eliminated as a ligand for the reasons given in Section IIA1. Arginine is absent in the plastocyanins. Tyrosine has been eliminated by optical absorption studies of azurin and by a recent analysis of the resonance enhanced Raman spectrum of stellacyanin (206).

Possible liganding groups which cannot be excluded on the basis of chemical studies are: carboxylate, peptide nitrogen, a terminal amino group, lysine, histidine methionine and hydroxyl groups. The ENDOR study of *Rist et al.* (28) on stellacyanin is the only evidence which definitely shows the presence of nitrogenous ligands to the copper. It appears that biochemists will have to await the results of crystallographic studies presently underway in Prof. L. Jensen's laboratory.

Type 2. This Cu^{2+} complex is characteristic of the multi-copper oxidases and is typified by its lack of a sufficient optical absorption to be observed above that of Type 1 Cu^{2+} and by its larger hyperfine coupling constant, A_{11}. It is often referred to as the more normal Cu^{2+} but this is quite misleading because instead of its physical properties being perturbed by its binding site, it has unusual chemical properties.

The best example of its special chemical properties is its possession of two readily occupied coordination positions which assume in-plane character as evidenced by the binding of two (weak field) F^- ions as in-plane ligands. This capability may be important in the mechanism of these enzymes.

EPR measurements have shown that this Cu^{2+} is coordinated at least in part to nitrogen atoms of the protein. Nothing is known about the liganding amino acids.

Type 2 Cu^{2+} shows considerable variation in its redox behavior, and only recently has it been shown to undergo oxidation-reduction in a kinetically competent fashion in *Polyporous laccase*.

Type 3. Least is known about this form of copper. The original proposal (*61*) that the Type 3 copper consisted of $Cu^{2+}Cu^{2+}$ pairs with strongly antiferromagnetically coupled spins is consistent with all subsequent observations, and this structure does offer a structural explanation for a 2-electron redox center functioning at rather high reduction potentials.

The possible similarity of this Cu—Cu pair to that observed in tyrosinase (*207*) and hemocyanin (*208*) deserves consideration, but the overall functional differences are evident. The most obvious difference between Type 3 Cu and the Cu binding sites of hemocyanin and tyrosinase concerns interaction with carbon monoxide. Laccases are not inhibited by CO (*113*) and therefore have a low or non-existent affinity for this molecule. In contrast, both tyrosinase and hemocyanin have a demonstrable affinity for CO. It is reasonable that CuCu pairs, like isolated Cu ions, are capable of a wide variety of structure-function combinations.

Acknowledgements. The author is deeply indebted to Professors *B. G. Malmström* and *T. Vänngård* for many helpful discussions and for their continued interest and support. Thanks go to Dr. *G. Pettersson* for a valuable communication, and Drs. *C. Briving* and *J. Deinum* for allowing their manuscript to be read in advance of publication. During the preparation of this review the author was supported by grants from the USPHS GM 18869 and 21519.

V. References

1. *Verhoeven, W., Takeda, T.*: In: The Johns Hopkins Press, p. 159—162 (*W. D. McElroy* and *B. Glass*, eds.) (1956).
2. *Horio, T.*: J. Biochem. (Tokyo) *45*, 195 (1958).
3. *Horio, T.*: J. Biochem. (Tokyo) *45*, 267 (1958).
4. *Sutherland, I. W., Wilkinson, J. F.*: J. Gen. Microbiol. *30*, 105 (1963).
5. *Sutherland, I. W.*: Arch. Mikrobiol. *54*, 350 (1966).
6. *Ambler, R. P., Brown, L. H.*: Biochem. J. *104*, 784 (1967).
7. *Tamanaka, T., Kijimoto, S., Okunuki, K.*: J. Biochem. (Tokyo) *53*, 256 (1963).
8. *Spies, J. R., Chambers, D. C.*: Anal. Chem. *20*, 30 (1948).
9. *Tang, S.-P., Coleman, J. E., Myer, Y. P.*: J. Biol. Chem. *243*, 4268 (1968).
10. *Maria, H. J.*: Nature *209*, 1023 (1966).
11. *Brill, A. S., Bryce, G. F., Maria, H. J.*: Biochim. Biophys. Acta *154*, 342 (1968).
12. *Fee, J. A., Malmström, B. G., Vänngård, T.*: Biochim. Biophys. Acta *197*, 136 (1970).
13. *Wetlaufer, D. B.*: Advan. Protein Chem. *17*, 303 (1962).
14. *Ambler, R. P., Brown, L. H.*: J. Mol. Biol. *9*, 825 (1964).
15. *Finazzi-Agro, A., Rotilio, G., Avigliano, L., Guerrieri, P., Boffi, V., Mondovi, B.*: Biochemistry *9*, 2009 (1970).
16. *KaLuk, G.*: Biopolymers *10*, 1229 (1971).
17. *Yagil, G.*: Tetrahedron *23*, 2855 (1967).
18. *Avigliano, L., Guerrieri, P., Calabrese, L., Vallogini, M. P., Rotilio, G., Mondovi, B., Finazzi-Agro, A.*: Ital. J. Biochem. *20*, 125 (1970).
19. *Finazzi-Agro, A., Giovagnoli, C., Avigliano, L., Rotilio, G., Mondovi, B.*: Eur. J. Biochem. *34*, 20 (1973).
20. *Koenig, S. H., Brown, R. D.*: Ann. N. Y. Acad. Sci. *222*, 752 (1973).

21. *Boden, N., Holmes, M. C., Knowles, P. F.*: Biochem. Biophys. Res. Commun. *57*, 845 (1974).
22. *Omura, T.*: Biochemistry *50*, 395 (1961).
23. *Stigbrand, T., Malmström, B. G., Vänngård, T.*: FEBS Letters *12*, 260 (1971).
24. *Peisach, J., Levin, W. G., Blumberg, W. E.*: J. Biol. Chem. *242*, 2847 (1967).
25. *Shichi, H., Hackett, D. P.*: Arch. Biochem. Biophys. *100*, 185 (1963).
26. *Stigbrand, T.*: Biochim. Biophys. Acta *236*, 246 (1971).
27. *Malmström, B. G., Reinhammar, B., Vänngård, T.*: Biochim. Biophys. Acta *205*, 48 (1970).
28. *Rist, G. H., Hyde, J. S., Vänngård, T.*: Proc. Natl. Acad. Sci. *67*, 79 (1970).
29. *Murpurgo, L., Finazzi-Agro, A., Rotilio, G., Mondovi, B.*: Biochim. Biophys. Acta *271*, 292 (1972).
30. *McMillin, D. R., Holwerda, R. A., Gray, H. B.*: Proc. Natl. Acad. Sci. *71*, 1339 (1974).
31. *Katoh, S.*: Nature *186*, 533 (1960).
32. *Gorman, D. S., Levin, R. P.*: Plant Physiol. *41*, 1637 (1966).
33. *Katoh, S., Shiratori, I., Takamiya, S.*: Biochemistry *51*, 32 (1962).
34. *Milne, P. R., Wells, J. R. E.*: J. Biol. Chem. *245*, 1566 (1970).
35. *Graziani, M. T., Finazzi-Agro, A., Rotilio, G., Barra, D., Mondovi, B.*: Biochemistry *13*, 804 (1974).
36. *Blumberg, W. E., Peisach, J.*: Biochim. Biophys. Acta *126*, 269 (1966).
37. *Katoh, S., Takamiya, A.*: Biochemistry *55*, 378 (1964).
38. *Malkin, R., Knaff, D. B., Bearden, A. J.*: Biochim. Biophys. Acta *305*, 675 (1973).
39. *Nakamura, T., Ogura, Y.*: Biochemistry *64*, 267 (1968).
40. *Pridham, J. B.* (ed.): Enzyme chemistry of phenolic compounds. New York: Macmillan Co. 1963.
41. *Yakushiji, E.*: Acta Phytochem. Japan *12*, 227 (1941).
42. *Gregg, D. C., Miller, W. H.*: J. Am. Chem. Soc. *62*, 1374 (1940).
43. *Benfield, G., Bocks, S. M., Bromley, K., Brown, B. R.*: Phytochemistry *3*, 79 (1964).
44. *Fahraeus. G., Ljunggren, H.*: Biochem. Biophys. Acta *46*, 22 (1961).
45. *Fahraeus, G.*: Biochem. Biophys. Acta *54*, 192 (1964).
46. *Bertrand, G.*: Compt. Rend. *118*, 1215 (1894).
47. *Keilen, D., Mann, T.*: Nature *145*, 304 (1940).
48. *Fahraeus, G., Tullander, V., Ljunggren, H.*: Physiol. Plantarum *11*, 631 (1958).
49. *Malmström, B. G., Fahraeus, G., Mosbach, R.*: Biochem. Biophys. Acta *28*, 652 (1958).
50. *Mosbach, R.*: Biochem. Biophys. Acta *73*, 204 (1963).
51. *Fahraeus, G., Reinhammer, B.*: Acta Chem. Scand. *21*, 2367 (1967).
52. *Jonsson, M., Pettersson, E., Reinhammer, B.*: Acta Chem. Scand. *22*, 2135 (1968).
53. *Butzow, J. J.*: Biochem. Biophys. Acta *168*, 490 (1968).
54. *Strothkamp, K. G., Dawson, C. R.*: Biochemistry *13*, 434 (1974).
55. *Briving, C., Deinum, J.*: FEBS Letters, *51*, 43 (1975).
56. *Malkin, R., Malmström, B. G.*: Advan. Enzymol. *33*, 177 (1970).
57. *Broman, L., Malmström, B. G., Aasa, R., Vänngård, T.*: J Mol. Biol. *5*, 3 (1962).
58. *Ehrenberg, A., Malmström, B. G., Broman, L., Mosbach, R.*: J Mol. Biol. *5*, 450 (1962).
59. *Malmström, B. G., Reinhammer, B., Vänngård, T.*: Biochem. Biophys. Acta *156*, 67 (1968).
60. *Vänngård, T.*: In: Magnetic resonance in biological systems (*A. Ehrenberg, B. G. Malmström* and *T. Vänngård*, (eds.), p. 213. New York—London: Pergamon Press 1967.
61. *Fee, J. A., Malkin, R., Malmström, B. G., Vänngård, T.*: J Biol. Chem. *244*, 4200 (1969).
62. *Malkin, R., Malmström, B. G., Vänngård, T.*: Eur. J. Biochem. *10*, 324 (1969).
63. *Fee, J. A., Malmström, B. G.*: Biochem. Biophys. Acta *153*, 299 (1968).
64. *Reinhammar, B. R. M.*: Biochem. Biophys. Acta *275*, 245 (1972).
65. *Malkin, R., Malmström, B. G., Vänngård, T.*: FEBS Letters *1*, 50 (1968).
66. *Chadwick, B. M., Sharpe, A. G.*: Advan. Inorg. Radiochem. *8*, 83 (1966).
67. *Malkin, R., Malmström, B. G., Vänngård, T.*: Eur. J. Biochem. *7*, 253 (1969).
68. *Bränden, R., Malmström, B. G., Vänngård, T.*: Eur. J. Biochem. *18*, 238 (1971).
69. *Bränden, R., Malmström, B. G., Vänngård, T.*: Eur. J. Biochem. *36*, 195 (1973).
70. *VanWart, H. E., Lewis, A., Scheraga, H. A., Saeva, F. D.*: Proc. Natl. Acad. Sci. *70*, 2619 (1973).
71. *Clark, W. M.*: Oxidation-reduction potentials of organic systems. Baltimore: Williams and Wilkins Co. 1960.

72. *Byers, W., Curzon, G., Garbett, K., Speyer, B. E., Young, S. N., Williams, R. J. P.:* Biochim. Biophys. Acta *310*, 38 (1973).
73. *Ginsberg, A. P.:* Inorg. Chim. Acta Rev. *5*, 545 (1971).
74. *Bossa, F., Rotilio, G., Fasella, P., Malmström, B. G.:* Eur. J. Biochem. *10*, 395 (1969).
75. *Malmström, B. G., Aasa, R., Vänngård, T.:* Biochim. Biophys. Acta *110*, 431 (1965).
76. *Malmström, B. G.:* Metalloenzymes Conference, Oxford *1*, 47 (1973).
77. *Nakamura, T.:* Biochim. Biophys. Acta *30*, 538 (1958).
78. *Makino, N., Ogura, Y.:* Biochemistry *69*, 91 (1971).
79. *Reinhammar, B. R. M., Vänngård, T. I.:* Eur. J. Biochem. *18*, 463 (1971).
80. *Kubowitz, F.:* Biochem. Z. *299*, 32 (1938).
81. *Tissieres, A.:* Nature *162*, 340 (1948).
82. *Omura, T.:* Biochemistry *50*, 390 (1961).
83. *Iwasaki, H., Matsubara, T., Mori, T.:* J. Biochem. *61*, 814 (1967).
84. *Ando, K.:* Biochemistry *68*, 501 (1970).
85. *Gregory, R. P. F., Bendall, D. S.:* Biochemistry *101*, 569 (1966).
86. *Nakamura, T.:* Biochem. Biophys. Res. Commun. *2*, 111 (1960).
87. *Nakamura, T.:* In: Free radicals in biological systems (*M. S. Blois*, Jr., ed.). New York: Academic Press 1961.
88. *Broman, L., Malmström, B. G., Aasa, R., Vänngård, T.:* Biochim. Biophys. Acta *75*, 365 (1963).
89. *Malmström, B. G.:* Pure Appl. Chem. *24*, 393 (1970).
90. *Andreasson, L.-E., Bränden, R., Malmström, B. G., Strömberg, C., Vänngård, T.:* In: Oxidases and related systems (*T. E. King, H. S. Mason* and *M. Morrison*, eds.). Baltimore: University Park Press 1973.
91. *Malmström, B. G.:* In: *J. A. Drenth, R. A. Oosterbaan*, and *C. Veeger* (eds.), Enzymes: Structure and Function, Proc. of the 8th Meeting of the Fed. Eur. Biochem. Soc., p. 119 (1972).
92. *George, P.:* In: Oxidases and Related redox systems (*T. E. King, H. S. Mason* and *M. Morrison*, eds.), p. 3. New York: John Wiley 1965.
93. *Pecht, I.:* Abstracts from Sixth FEBS Meeting, Madrid, p. 150 (1969).
94. *Malmström, B. G., Finazzi-Agro, A., Antonini, E.:* Eur. J. Biochem. *9*, 383 (1969).
95. *Andreasson, L.-E., Malmström, B. G., Strömberg, C., Vänngård, T.:* Eur. J. Biochem. *34*, 434 (1973).
96. *Andreasson, L.-E., Branden, R., Malström, B. G., Vänngård, T.:* FEBS Letters *32*, 187 (1973).
97. *Holwerda, R. A., Gray, H. B.:* J. Am. Chem. Soc. *96*, 6008 (1974).
98. *Branden, R., Reinhammar, B.:* unpublished results.
99. *Dawson, J. W., Gray, H. B., Holwerda, R. A., Westhead, E. W.:* Proc. Natl. Acad. Sci. *69*, 30 (1972).
100. *Holmberg, C. G.:* Acta Physiol. Scand. *8*, 227 (1944).
101. *Holmberg, C. G., Laurell, C. B.:* Acta Chem. Scand. *2*, 550 (1948).
102. *Holmberg, C. G., Laurell, C. B.:* Acta Chem. Scand. *5*, 476 (1951).
103. *Osaki, S., Johnson, D. A., Frieden, E.:* J. Biol. Chem. *241*, 2746 (1966).
104. *Shokeir, M. H., Shreffler, D. C.:* Proc. Natl. Acad. Sci. *62*, 867 (1969).
105. *Marceau, N., Aspin, N.:* Biochim. Biophys. Acta *293*, 338 (1973).
106. *Marceau, N., Aspin, N.:* Biochim. Biophys. Acta *328*, 351 (1973).
107. *Shokeir, M. H. K.:* Clin. Biochem. *5*, 115 (1972).
108. *Sternlieb, I., Scheinberg, I. H.:* Ann. N. Y. Acad. Sci. *94*, 71 (1961).
109. *Broman, L.:* Nature, *182*, 1655 (1958).
110. *Hirschman, S. Z., Morell, A. G., Scheinberg, I. H.:* Ann. N. Y. Acad. Sci. *94*, 960 (1961).
111. *Deutsch, H. F., Fisher, G. B.:* J. Biol. Chem. *239*, 3325 (1964).
112. *Ryden, L.:* Intern. J. Protein Res. *III*, 131 (1971).
113. *Ryden, L.:* Intern. J. Protein Res. *III*, 191 (1971).
114. *Kaya, T.:* Biochemistry *56*, 122 (1964).
115. *Poillon, W. N., Bearn, A. G.:* Biochim. Biophys. Acta *127*, 407 (1966).
116. *Richterich, R., Temperli, A., Aebi, H.:* Biochim. Biophys. Acta *56*, 240 (1962).

117. *Mukasa, H., Kajiyama, S., Sugiyama, K., Funakubo, K., Itah, M., Nosoh, Y., Sato, T.:* Biochim. Biophys. Acta *168*, 132 (1968).
118. *Vasilets, I. M., Konnova, L. A., Kushner, V. P., Bozhkov, V. M., Zdrodovskaya, E. P., Mukha, G. V.:* Biokhimiya, *33*, 1285 (1968).
119. *Ryden, L.:* Acta. Universitatis Upsaliensis, Abstracts of Uppsala Dissertations from the Faculty of Science *222* (1972).
120. *Ryden, L.:* FEBS Letters *18*, 321 (1971).
121. *Ryden, L.:* Eur. J. Biochem. *28*, 46 (1972).
122. *Ryden, L.:* Eur. J. Biochem. *26*, 380 (1972).
123. *Deinum, J., Vänngård, T.:* Biochim. Biophys. Acta *310*, 321 (1973).
124. *Kasper, C. B., Deutsch, H. F.:* J. Biol. Chem. *238*, 2325 (1963).
125. *Hibno, Y., Samejima, T.:* Arch. Biochem. Biophys. *130*, 617 (1969).
126. *Witwicki, J., Zakrzewski, K.:* Eur. J. Biochem. *10*, 284 (1969).
127. *Bannister, W. H., Wood, E. J.:* Eur. J. Biochem. *11*, 179 (1969).
128. *Nylen, U., Pettersson, G.:* Eur. J. Biochem. *27*, 578 (1972).
129. *Jamieson, G. A.:* J. Biol. Chem. *240*, 2019 (1965).
130. *Morell, A. G., Van Den Hamer, C. J. A., Scheinberg, I. H., Ashwell, G.:* J. Biol. Chem. *241*, 3745 (1966).
131. *Morell, A. G., Irvine, R. A., Sternlieb, I., Scheinberg, I. H.:* J. Biol. Chem. *243*, 155 (1968).
132. *Gregoriadis, G., Morell, A. G., Sternlieb, I., Scheinberg, I. H.:* J. Biol. Chem. *245*, 5833 (1970).
133. *Magdoff-Fairchild, B., Lovell, F. M., Low, B. W.:* J. Biol. Chem. *244*, 3497 (1969).
134. *Morell, A. G., Van Den Hamer, C. J. A., Scheinberg, I. H.:* J. Biol. Chem. *244*, 3494 (1969).
135. *Aisen, P., Koenig, S. H., Lilienthal, H. R.:* J. Mol. Biol. *28*, 225 (1967).
136. *Deutsch, H. F.:* Arch. Biochem. Biophys. *89*, 225 (1960).
137. *Deutsch, H. F., Kasper, C. B., Walsh, D. A.:* Arch. Biochem. Biophys. *99*, 132 (1962).
138. *Nakagawa, O.:* Intern. J. Peptide Protein Res. *4*, 385 (1972).
139. *Andreasson, L.-E., Vänngård, T.:* Biochim. Biophys. Acta *200*, 247 (1970).
140. *Veldsema, A., VanGelder, B. F.:* Biochim. Biophys. Acta *293*, 322 (1973).
141. *Gunnarsson, P.-O., Nylén, U., Pettersson, G.:* Eur. J. Biochem. *37*, 47 (1973).
142. *Van Leeuwen, F. X. R., Wever, R., Van Gelder, B. F.:* Biochem. Biophys. Acta, *315*, 200 (1973).
143. *Blumberg, W. E., Eisinger, J., Aisen, P., Morell, A. G., Scheinberg, I. H.:* J. Biol. Chem. *238*, 1675 (1963).
144. *Kasper, C. B.:* J. Biol. Chem. *243*, 3218 (1968).
145. *Wever, R., Van Leeuwen, F. X. R., Van Gelder, B. F.:* Biochim. Biophys. Acta *302*, 236 (1973).
146. *Laurell, C. B.:* In: The plasma proteins (*F. W. Putnam*, ed.), p. 349. New York: Academic Press 1966.
147. *Van Gelder, B. F., Veldsema, A.:* Biochim. Biophys. Acta *130*, 267 (1966).
148. *Curzon, G.:* Biochem. J. *77*, 66 (1960).
149. *Boyd, D. B.:* J. Am. Chem. Soc. *94*, 8799 (1972).
150. *Neubert, L. A., Carmack, M.:* J. Am. Chem. Soc. *96*, 943 (1974).
151. *Walaas, E., Lovstad, A., Walaas, O.:* R. Arch. Biochem. Biophys. *121*, 480 (1967).
152. *Lovstad, R. A.:* Eur. J. Biochem. *8*, 303 (1969).
153. *Curzon, G.:* Biochem. J. *103*, 289 (1967).
154. *Sekiguchi, T., Nosoh, Y.:* Arch. Biochem. Biophys. *138*, 319 (1970).
155. *Pettersson, G., Pettersson, I.:* Acta Chem. Scand. 23 3235 (1969).
156. *Pettersson, G.:* Acta Chem. Scad. *24*, 1809 (1970).
157. *Pettersson, G.:* Acta Chem. Scand. *22*, 3063 (1968).
158. *Paulsson, L.-E., Pettersson, G.:* Acta Chem. Scand. *23*, 2727 (1969).
159. *Pettersson, G.:* Acta Chem. Scand. *23*, 2317 (1969).
160. *Pettersson, G., Pettersson, I.:* Acta Chem. Scand. *24*, 1275 (1970).
161. *Curzon, G., O'Reilly, S.:* Biochem. Biophys. Res. Commun. *2*, 284 (1960).
162. *Levine, W. G., Peisach, J.:* Biochim. Biophys. Acta *77*, 602 (1963).

163. *McDermott, J. A., Huber, C. T., Osaki, S., Frieden, E.:* Biochim. Biophys. Acta *151,* 541 (1968).
164. *Morell, A. G., Aisen, P., Scheinberg, I. H.:* J. Biol. Chem. *237* 3455 (1962).
165. *Curzon, G.:* Biochem. J. *79,* 656 (1961).
166. *Gunnarsson, P.-O., Nylen, U., Pettersson, G.:* Eur. J. Biochem. *37,* 41 (1973).
167. *Young, S. N., Curzon, G.:* Biochem, J. *129,* 273 (1972).
168. *Frieden, E., Osaki, S., Kobayashi, H.:* J. Gen Physiol. *49,* 213 (1965).
169. *Osaki, S.:* J. Biol. Chem. *241,* 5053 (1966).
170. *McKee, D. J., Frieden, E.:* Biochemistry *10,* 3880 (1971).
171. *Osaki, S., Walaas, O.:* Arch. Biochem. Biophys. *125,* 918 (1968).
172. *Carrico, R. J., Malmström, B. G., Vänngård, T.:* Eur. J. Biochem. *22,* 127 (1971).
173. *Manabe, T., Hatano, H., Hiromi, K.:* J. Biochem. (Tokyo) *73,* 1169 (1973).
174. *Levine, W. G., Peisach, J.:* Biochim. Biophys. Acta *63,* 528 (1962).
175. *Curzon, G., Speyer, B. E.:* Biochem. J. *105,* 243 (1967).
176. *Pettersson, G.:* Acta Chem. Scand. *24,* 1838 (1970).
177. *Gunnarsson, P.-O., Pettersson, G., Pettersson, I.:* Eur. J. Biochem. *17,* 586 (1970).
178. *Streitweiser, A.,* Jr.: Molecular orbital theory for organic chemists, Chap. 7. New York: John Wiley 1961.
179. *Manabe, T., Manabe, N., Hiromi, K., Hatano, H.:* FEBS Letters *23,* 268 (1972).
180. *Osaki, S.:* Unpublished results.
181. *Osaki, S., Walaas, O.:* J. Biol. Chem. *242,* 2653 (1967).
182. *Curzon, G.:* Biochem. J. *100,* 295 (1966).
183. *Gunnarsson, P.-O., Nylén, U., Pettersson, G.:* Eur. J. Biochem. *27,* 572 (1972).
184. *Manabe, T., Manabe, N., Hiromi, K., Hatano, H.:* FEBS Letters *16,* 201 (1971).
185. *Speyer, B. E., Curzon, G.:* Biochem. J. *106,* 905 (1968).
186. *Curzon, G., Speyer, B. E.:* Biochem. J. *190,* 25 (1968).
187. *Imoto, T., Akasaka, K., Hatano, H.:* FEBS Letters *15,* 149 (1971).
188. *Gunnarsson, P.-O., Pettersson, G.:* Eur. J. Biochem. *27,* 564 (1972).
189. *Stark, G. R., Dawson, C. R.:* In: The enzymes (*P. D. Boyer, H. A. Lardy* and *K. Myrbäck,* eds.), Vol. VIII, p. 297. New York: Academic Press 1963.
190. *Dawson, C. R.:* In: The biochemistry of copper (*J. Peisach, P. Aisen* and *W. E. Blumberg,* eds.), New York: Academic Press 1966.
191. *Nakamura, T., Makina, N., Ogura, Y.:* Biochemistry *64,* 188 (1968).
192. *Lee, M. H., Dawson, C. R.:* J. Biol. Chem. *248,* 6596 (1973).
193. *Lee, M. H., Dawson, C. R.:* J. Biol. Chem. *248,* 6603 (1973).
194. *Yamazaki, I., Piette, L. H.:* Biochim. Biophys. Acta *50,* 62 (1961).
195. *Morell, A. G., Scheinberg, I. H.:* Science, *127,* 588 (1958).
196. *Aisen, P., Morell, A. G.:* J. Biol. Chem. *240,* 1974 (1965).
197. *Morell, A. G., Aisen, P., Blumberg, W. E., Scheinberg, I. H.:* J. Biol. Chem. *239,* 1042 (1964).
198. *Kasper, C. B., Deutsch, H. F., Beinert, H.:* J. Biol. Chem. *238,* 2338 (1963).
199. *Curzon, G., Vallet, L.:* Biochem. J. *74,* 279 (1959).
200. *Kasper, C. B.:* Biochemistry *6,* 3185 (1967).
201. *Malmström, B. G., Vänngård, T.:* J. Mol. Biol.*2,* 118 (1960).
202. *Mason, H. S.:* Biochem. Biophys. Res. Commun. *10,* 11 (1963).
203. *Blumberg, W. E.:* In: The biochemistry of copper (*J. Peisach, P. Aisen* and *W. E. Blumberg,* eds.), p. 49. New York: Academic Press 1966.
204. *Brill, A. S., Bryce, G. F.:* J. Chem. Phys. *48,* 4398 (1968).
205. *Vänngård, T.:* In: Biological applications of electron spin resonance, p. 411. New York: Wiley-Interscience 1972.
206. *Siiman, O., Young, N. M., Carey, P. R.:* J. Am. Chem. Soc. *96,* 5583 (1974).
207. *Makino, N., McMahill, P., Mason, H. S., Moss, T. H.:* J. Biol. Chem. *249,* 6062 (1974).
208. *Schoot Uiterkamp, A. J. M.:* FEBS Letters *20,* 93 (1972).
209. *Horio, T., Sekuzu, I., Higashi, T., Okunuki, K.:* In: Haematin enzymes, I.U.B. Symposium 19, p. 302 (1959).
210. *Coval, M. L., Horio, T., Kamen, M. E.:* Biochim. Biophys. Acta *51,* 246 (1961).
211. *Suzuki, H., Iwasaki, H.:* J. Biochem. (Tokyo) *52,* 193 (1962).

212. *Strahs, G.:* Science *165,* 60 (1069).
213. *Ambler, R. P.:* Biochem. J. *89,* 341 (1963).
214. *Reinhammer, B.:* Biochim. Biophys. Acta *205,* 35 (1970).
215. *Falk, K.-E., Reinhammar, B.:* Biochim. Biophys. Acta *285,* 84 (1972).
216. *Paul, K.-G., Stigbrand, T.:* Biochim. Biophys. Acta *221,* 255 (1970).
217. *Stigbrand, T.:* FEBS Letters *23,* 41 (1972).
218. *Stigbrand, T., Sjöholm, I.:* Biochim. Biophys. Acta *263,* 244 (1972).
219. *Ida, S., Morita, Y.:* Agri. Biol. Chem. (Tokyo) *33,* 10 (1969).
220. *Nakamura, T.:* Biochim. Biophys. Acta *50,* 44 (1958).
221. *Omura, T.:* J. Biochem. (Tokyo) *50,* 265 (1961).
222. *Stark, G. R., Dawson, C. R.:* J. Biol. Chem. *237,* 712 (1962).
223. *Kelly, J., Ambler, R. P.:* Trans. Biochem. Soc. *1,* 164 (1973).
224. *Ramshaw, J. A. M., Scaswen, M. D., Bailey, C. J., Boulter, D.:* Biochem. J. *139,* 583 (1974).
225. *Deinum, J., Reinhammar, B., Marchesini, A.:* FEBS Letters *42,* 241 (1974).
226. *Freeman, S., Daniel, E.:* Biochemistry *12,* 4806 (1973).
227. *Levine, W.G., Peisach J.:* Nature 207 406 (1965).

Received January 7, 1975

Mechanisms of Zinc Ion Catalysis in Small Molecules and Enzymes

Michael F. Dunn

Department of Biochemistry, University of California, Riverside, California 92502

Table of Contents

I. Introduction

It has been the intention of the author in this review to examine the roles played by zinc ion in homogeneous solution catalysis both for small molecule-zinc ion complexes and for zinc-metalloenzymes. Emphasis is placed on the integration of physical-inorganic mechanistic concepts derived from studies on small molecule systems with the accumulated kinetic, chemical, and structural information available on select enzyme examples in order that reasonable mechanistic hypotheses might be developed for the roles played by zinc ion in enzymatic catalysis.

The small molecule examples discussed in this review provide a rich collection of data pertaining to the catalytic functions which zinc ion can perform as an electrophile in aqueous solution. The high resolution structural information on the zinc-metalloproteins together with information derived from chemical modifications, kinetic studies, and other spectroscopic techniques place severe constraints on the possible roles played by zinc ion in enzyme catalytic mechanism. As a consequence, the focus of this review is placed on the relationship of these catalytic functions, as defined by the small molecule examples, to the probable roles played by zinc ion in enzyme catalysis.

Because the scope of this review encompasses a variety of topics, it has been necessary to restrict detailed discussion to those aspects of each topic which are germane to the treatment of mechanism. For this reason, the emphasis of the review is placed on developments which occurred during the last five to ten years, and no attempt is made to treat each topic in full historical detail.

II. Structural and Chemical Properties of Zn(II) Complexes

Comprehensive reviews of divalent metal ion coordination chemistry and kinetic behavior can be found in standard references such as *Cotton* and *Wilkinson* (*1*) and the works of *Eigen* and *Wilkens* (*2—5*). Therefore, discussion of zinc coordination complexes here is restricted to consideration of those aspects of zinc chemistry which relate to the catalytic properties of Zn-metalloenzymes.

1. Electronic Structure

Divalent zinc ion is characterized by a completely filled shell of d electrons inside a vacant s shell. This relatively stable electronic configuration accounts for the ubiquity of the divalent oxidation state; the trivalent state is unknown, and although evidence for the univalent state exists, there are no examples of stable univalent zinc compounds (*1*). Because the d shell is completely filled, the stereochemistry and coordination number exhibited by Zn(II) complexes are specified solely by considerations of ligand size, electrostatic forces and covalent bonding forces. *Zn(II) complexes are not subject to ligand field stabilization interactions.* As a consequence, coordination numbers for zinc ion ranging from 2 to 8 have been reported, although coordination numbers of 4, 5 and 6 are most common (*1*). Examples of both tetrahedral and square planar geometries have been found for complexes of coordination number four. Regular or warped tetrahedral geometries are especially common. Typical structures for some four- and five-coordinate complexes are shown in Fig. 1. Five-coordinate complexes assume either a trigonal-bipyramid, or a square-pyramid geometry (Fig. 1 B). Small, electrically neutral ligands (*e.g.*, H_2O or NH_3) form octahedral complexes with zinc (*1, 8*).

2. Kinetics and Mechanism of Ligand Substitution

The equilibrium constant (K_D) for the dissociation of the nth water molecule from the zinc aquo ion [Eq. (1)]

$$\text{[Zn(OH}_2)_n]^{2+} \underset{}{\overset{K_D}{\rightleftharpoons}} \text{[Zn(OH}_2)_{n-1}]^{2+} + H_2O \tag{1}$$

$$K_D = k_r/k_t = (\text{[Zn(OH}_2)_{n-1}]^{2+}) (H_2O)/(\text{[Zn(OH}_2)_n]^{2+})$$

is approximately 10^{-3}M. *Eigen* (*3*) has reported that the specific first order rate constant, k_t, for dissociation is $\sim 4 \times 10^7 \text{ sec}^{-1}$. The magnitudes of k_t and K_D

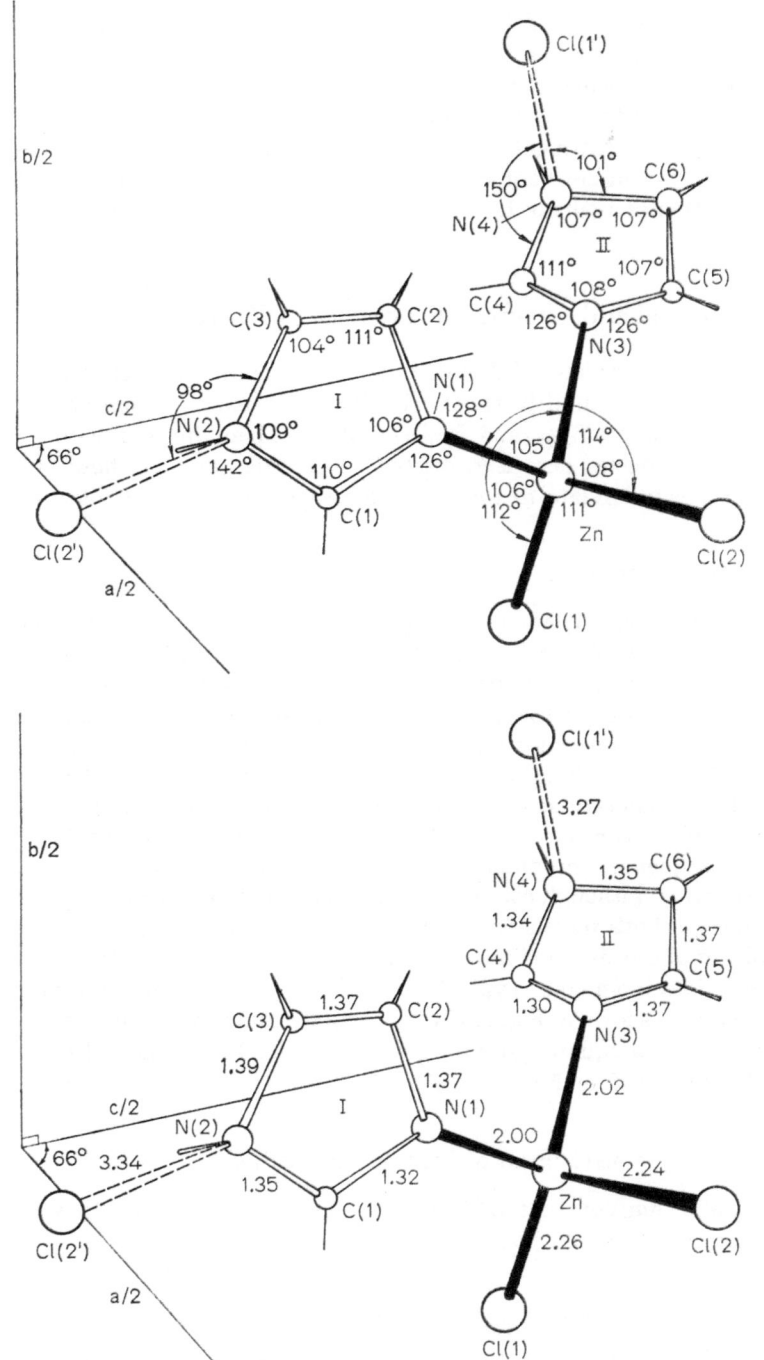

Fig. 1. A. The molecular structure of the tetrahedral complex, Zn(imidazole)$_2$Cl$_2$, determined from X-ray diffraction studies. Bond lengths and bond angles are shown. Taken from Ref. (6) with permission.

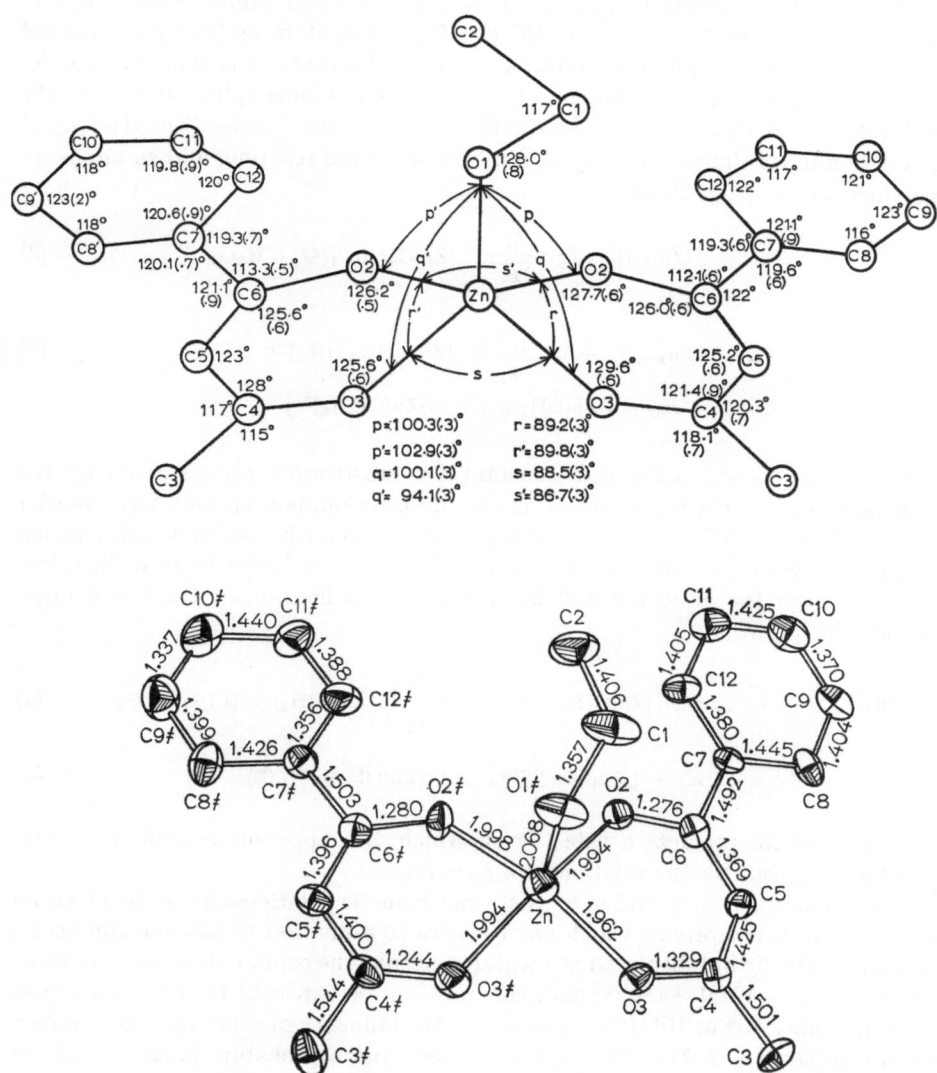

Fig. 1 B. The molecular structure of the five-coordinate square pyramidal complex, bis(benzoylacetonato)zinc monoethanolate. Bond lengths and bond angles are shown. Taken from Ref. (7) with permission

indicate that both k_f and k_r are essentially diffusion limited processes ($k_r = k_f/K_D$ $\simeq 10^{10} M^{-1}$ sec^{-1}).

The apparent second-order rate constants measured for substitution of inner-sphere H_2O by other ligands invariably show little or no dependence on ligand structure. Thus, acetate, sulfate, and chloride ions all combine with the aquo-zinc complex with $k_2 \simeq 3 \times 10^7 M^{-1}$ sec^{-1} at 25° (2), while complex formation with the neutral tridentate ligand pyridine-2-azodimethyl-aniline gives a slightly lower value, $4 \times 10^6 M^{-1} sec^{-1}$, at $15°$ (9). *Eigen* and *Wilkens* (2) have proposed that the only reasonable interpretation of these observations is that the kinetics of ligand substitution are dominated by dissociation of inner sphere H_2O from the metal aquo ion. Since complex formation between the species $[Zn(OH_2)_{n-1}]^{2+}$ and ligand must almost certainly be a diffusion limited reaction, and assuming an $S_N 1$ mechanism [Eq. (2, 3)],

$$[Zn(OH_2)_n]^{2+} \underset{k_r}{\overset{k_f}{\rightleftharpoons}} [Zn(OH_2)_{n-1}]^{2+} + H_2O \qquad (2)$$

$$[Zn(OH_2)_{n-1}]^{2+} + L \underset{k_L}{\overset{k_L}{\rightleftharpoons}} [Zn(OH_2)_{n-1}(L)]^{2+} \qquad (3)$$

$$K_D = ([Zn(OH_2)_{n-1}]^{2+})/([Zn(OH_2)_n]^{2+})$$

the apparent second order rate constant for substitution (k_2) is given by the relationship, $k_2 = K_D k_L$, where k_L is the diffusion-limited, specific second-order rate constant $(\sim 10^{10} M^{-1} sec^{-1})$ for complex formation. Alternatively, substitution could occur *via* rapid outer sphere complex formation followed by slow, $S_N 1$, loss of water from the aquo ion and inner sphere coordination of the substituting ligand [Eq. (4, 5)].

$$[Zn(OH_2)_n]^{2+} + L \overset{K_L}{\rightleftharpoons} [Zn(OH_2)_n]^{2+} \cdots L \underset{k_i}{\overset{k_i}{\rightleftharpoons}} [Zn(OH_2)_{n-1}(L)]^{2+} + H_2O \qquad (4)$$

$$K_L = ([Zn(OH_2)_n]^{2+} \cdots L)/([Zn(OH_2)_n]^{2+}) (L) \qquad (5)$$

This mechanism predicts a rate law in which the apparent second-order rate constant is given by the relationship, $k_2 = K_L k_1$.

The consequence of either of these mechanistic relationships is to place an upper limit on the rate at which inner sphere coordination to zinc ion can occur. So long as the ligand displaced is a water molecule, the combination rate can be no greater than $\sim 10^7 M^{-1} sec^{-1}$; and, this rate can be expected to show some variation (possibly 10^6 to $10^8 M^{-1} sec^{-1}$) due to the influence of other ligands in mixed aquo complexes (10, 11). Although water desorption probably limits the rate of enzyme-substrate combination for some zinc enzymes (viz. liver alcohol dehydrogenase, see Section IV—2), the rate of substrate turnover is not likely to be affected, since subsequent steps in the overall chemical transformation and/or product desorption invariably are rate limiting. Enzyme-substrate turnover numbers as a rule are 10^2 to 10^4 [substrate converted]/[site]/sec, *Eigen* and *Hammes* (4), *Bernhard* (12). Although carbonic anhydrase is a notable exception to this rule (see Section IV—3), this circumstance generally holds for zinc-enzyme systems.

The insensitivity of the formation rate constants to the nature of the ligand is ubiquitous for non-transition and transition metal ions alike. This phenomenon is believed to be the result of either an $S_N 1$ mechanism [Eq. (2, 3)], or a two-step, $S_N 1$-like mechanism [Eq. (4, 5)] involving first outer sphere complex formation followed by $S_N 1$ loss of water from the aquo ion (1, 2).

3. Comparison of Zinc Ion to Other Divalent Metal Ions

Since some of the evidence bearing on the mechanistic role played by zinc ion in the enzyme examples discussed in this review is based on the properties of enzymes in which another divalent metal ion has been substituted for zinc ion, it is of interest here to briefly consider some of the properties of the ions which are commonly utilized in substitution studies. These metal ions are Ni^{2+}, Co^{2+}, Fe^{2+}, Mn^{2+}, Cu^{2+}, Cd^{2+}, and Hg^{2+}.

The first five ions are members of the d transition elements. Their divalent oxidation states are characterized by partially filled d shells. Therefore, these metal ions have UV-visible spectral properties and magnetic properties which are of use in probing catalytic mechanism. These transition elements all exhibit multiple, relatively, stable oxidation states. All are capable of forming (relatively stable) 4—6 coordinate ligand complexes. Unlike Zn^{2+} (and Cd^{2+}, Hg^{2+}) ligand field stabilization has a strong influence on the properties (both structural and chemical) of the ligand complexes of these ions. Cd and Hg are grouped together with Zn in the periodic table since they are electronically homologous. However, the chemistry of Hg is quite different in several respects from that of Cd and Zn. For example, although higher (4 to 6) coordinate states are known for Hg(II), this ion exhibits a propensity for the formation of essentially covalent, two-coordinate complexes.

Table 1, compiled from data given in *Cotton* and *Wilkinson* (1) and from *Eigen* and *Wilkens* (2, 3), illustrates the differences in inner sphere water lability, as measured by H_2O substitution rates, for the aquo ions. Note that within a series (Ni—Cu and Zn—Hg) there is an approximate correlation between ionic radius and the lability of bound water. *Eigen* and *Wilkens* (2) have suggested that the deviation from this trend shown by Cu(II) is the result of a Jahn-Teller effect which results in the distortion of the octahedral complex and the labilization of the apical ligands.

Table 1. Selected properties of divalent metal ion aquo complexes

Divalent metal ion	Ionic radii (Å)[a]		k_f[a] (sec^{-1})	pK_a'[b]
	Goldschmidt	*Pauling*		
Ni^{2+}	0.68	0.69	2×10^4	10.18 ± 0.1
Co^{2+}	0.82	0.74	5×10^5	9.85 ± 0.05
Fe^{2+}	0.83	0.76	2×10^6	7.15 ± 0.2
Mn^{2+}	0.91	0.80	7×10^6	10.76 ± 0.04
Cu^{2+}	0.72	—	5×10^8	$\sim 7.3 \pm 0.4$
Zn^{2+}	0.69	0.74	3×10^7	9.13 ± 0.06
Cd^{2+}	1.03	0.97	4×10^8	9.0 ± 0.2
Hg^{2+}	0.93	1.10	2×10^9	3.68 ± 0.2

[a] Values taken from Refs. (1) and (3). The values of K refer to the rate of dissociation of the nth water molecule from the aquo ion complex, as in Eq. (1).

[b] Values taken from Table 2 in Ref. (14). pK_a' values refer to the ionization process described by Eq. (6), see Text.

Zeltmann and *Morgan* (*13*) have shown that Co(II) mixed aquo complexes undergo H_2O exchange at rates which are dependent on the coordination state of the metal ion. They find that the rate of H_2O exchange is greater for tetrahedral complexes than for octahedral complexes by at least an order of magnitude. Exchange *via* an S_N1 mechanism [Eqs. (2—4)] would be expected to occur more slowly in tetrahedral complexes. Therefore, the observed dependence of rate on the coordination state suggests a change in mechanism, and these authors propose that ligand exchange occurs *via* an addition-elimination mechanism involving penta-coordinate species in the tetrahedral Co(II) complexes.

The pK_a' values [taken from *Sillén* and *Martell* (*14*) and references cited therein] for the ionization of the metal ion aquo-complex [Eq. (6)]

$$[M(H_2O)_n]^{2+} \underset{}{\overset{K_a}{\rightleftharpoons}} [M(H_2O)_{n-1}(OH)]^+ + H^+ \tag{6}$$

$$K_a = (H^+)\,([M(H_2O)_{n-1}(OH)]^+)/([M(H_2O)_n]^{2+})$$

are also listed in Table 1. These values are of particular interest since (as will become evident in the following sections) an inner sphere coordinated H_2O molecule may be involved either as a general acid-base catalyst or as a nucleophile in one or more of the various enzyme mechanisms under discussion.

III. Zinc Ion Catalysis in Model Enzyme Studies

The literature relevant to model systems involving electrophilic catalysis by divalent metal ions suggests the following catalytic functions for the involvement of zinc ion in enzymic catalysis:

(i) Lewis acid catalysis involving activation of substrate chemical bonds by bond polarization *via* inner sphere coordination to zinc ion.

(ii) Enhancement of ligand nucleophilicity as a consequence of pK_a' perturbation by coordination to zinc ion.

(iii) Facilitation of nucleophilic attack on substrate by charge neutralization in the ground and/or transition states.

(iv) Facilitation of reaction by the precise alignment of reactants at the enzyme site through coordination to zinc ion as a template.

Lewis (general) acid catalysis can exert an influence on the activation energy for reaction by activation of the ground-state by polarization of chemical bonds. Catalysis *via* bond polarization will be most effective if stabilization of the transition state both by bond polarization and by favorable electrostatic interaction between the metal ion and a developing negative charge on the substrate can occur. Coulombic interaction between nucleophile and metal ion can result in a large perturbation of the pK_a' for an inner sphere-coordinated nucleophile. For zinc ion the perturbation can be as large as 3—4 pK_a' units for —OH and —NH nucleophiles (see Table 1). Coulombic interactions between the metal ion and a negatively charged substrate can decrease repulsive charge-charge (and charge-dipole) forces which otherwise hinder the attack of an external nucleophile. The precise prealignment of reactants *via* coordination to the metal ion may be manifest as a large advantage in the entropy of activation (as much as 15—50 eu) (*15, 16*).

The transition state for an uncatalyzed, second-order (or higher-order) reaction usually corresponds to the most entropically unfavorable arrangement of the reacting species on the reaction coordinate, since the randomly-oriented reactants must be brought together from dilute solution into a highly-oriented, activated complex. However, if the catalyzed pathway involves a rate-limiting chemical reaction that proceeds from a pre-equilibrium complex in which the reactants and the catalyst have been brought together in the proper alignment for reaction (*15—17*), this disadvantage can be shifted to a non-rate-limiting step. The available evidence derived from both mechanism and structure studies (*15—17*) is highly supportive for a theory of enzyme catalysis in which these considerations make a significant contribution to the high efficiency of enzyme catalysis. The "intramolecular advantage" as here presented has been discussed in detail by *Bruice* (*15*) and by *Page* and *Jencks* (*16*). A somewhat different interpretation for the "intramolecular advantage" has been suggested by *Koshland et al.* (*18*).

For both model systems and metalloenzymes the role played by divalent metal ion is likely to involve a combination of the above functions. Indeed, as will be evident, the divalent metal ion clearly is involved in more than one function in at least four of the model systems discussed in the following paragraphs. These examples have been chosen because they lucidly illustrate these catalytic functions in reactions involving zinc ion. It is the conviction, or perhaps bias, of the author that these examples define in broad outline the full range of catalytic mechanisms involving electrophilic catalysis by zinc ion, both in enzymatic and nonenzymatic reactions.

1. Lewis Acid Catalysis

Breslow et al. (*19*) have investigated the mechanism of the divalent metal ion catalyzed hydration of 2-cyano-1,10-phenanthroline to the corresponding amide [see Eq. (7) below]. The cupric ion catalyzed reaction is extremely rapid ($t_{1/2}$ < 10 sec at 25°, pH 6—7). The Ni(II) and Zn(II) reactions, although slower, are greatly accelerated [the Ni(II) reaction by a factor of $\sim 10^7$] in comparison to the rate of hydration observed in the absence of divalent metal ion catalysis. The reaction obeys a rate law which is second-order over all: first-order with respect to hydroxide ion concentration, and first-order with respect to the 1:1 metal ion-substrate complex concentration. These authors suggest that the metal ion acts as a Lewis (general) acid in activating the nitrile for external nucleophilic attack by hydroxide ion, as illustrated in Eq. (7):

The kinetically equivalent mechanism involving attack by a coordinated hydroxyl was rejected by these authors because (a) reaction of 2-cyano-1,10-phenanthroline with the Zn(II) or Cu(II) monocomplexes of pyridine-2-carboxaldoxime, 2-aminomethyl-8-hydroxyquinoline, or the Cu(II) monocomplex of 2-hydroxy-1,10-phenanthroline gives the amide as the sole product. (b) The product of the Cu(II)-catalyzed reaction is not altered by the presence of ethylenediamine or hydroxyethylethylenediamine, although the rate of the reaction is inhibited. (c) The presence of high concentrations of either ethanol or ethanolamine

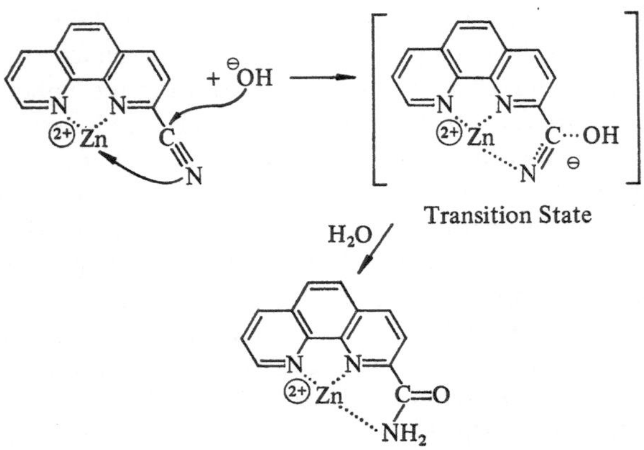

(7)

gives low yields of products derived from attack by these nucleophiles on the cupric complex. However, the proportion of the total reaction involving these nucleophiles is independent of the presence or absence of Cu(II).

These findings are clearly inconsistent with a mechanism involving nucleophilic attack *via* coordination of the nucleophile to the metal ion. The authors further argue that coordination of the nitrile in the ground-state is relatively insignificant due to the geometry of the complex and to the rigidity of the nitrile group, but that coordination in the transition-state is greatly favored. Therefore, the authors propose that the enormous rate accelerations found for this system stem from the distortion of the nitrile toward the geometry of the transition-state.

Zinc ion is essential for the catalytic activities of both yeast and liver alcohol dehydrogenase. Until recently, model systems have been notably unsuccessful in accounting for the participation of Zn(II) in the enzyme-catalyzed oxidoreductive interconversion of aldehyde and alcohol. The studies of *Creighton* and *Sigman* (20) and of *Shinkai* and *Bruice* (21, 22) conclusively demonstrate that Lewis (general) acid catalysis by Zn^{2+} (and other divalent metal ions) can effectively promote aldehyde reduction by the reduced 1,4-dihydropyridine moiety.

Creighton and *Sigman* (20) have investigated the zinc-ion-catalyzed reduction of 1,10-phenanthroline-2-carboxaldehyde by N-propyl-1,4-dihydronicotinamide. They found that in anhydrous acetonitrile, the zinc ion-1,10-phenanthroline-2-carboxaldehyde complex is rapidly reduced by the 1,4-dihydronicotinamide group [Eq. (8)]:

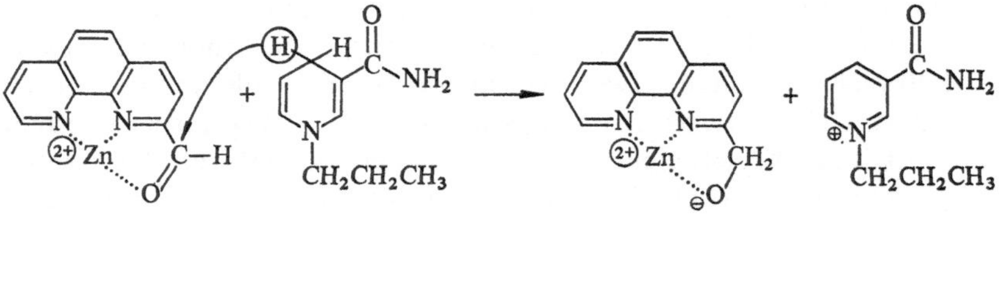

(8)

Since no detectable reaction occurs in the absence of zinc ion, the magnitude of the rate enhancement effected by Zn^{2+} must be enormous. Nevertheless, this facile reaction does not occur in protic milieu. Instead, in aqueous solution there occurs a rapid (virtually instantaneous) hydration of the carboxaldehyde carbonyl to the stable hemiacetal.

This reduction almost certainly proceeds *via* the activation of the aldehyde carbonyl both by bond polarization in the ground-state and by stabilization of bonding interactions in the transition state. Although deuterium isotope labeling experiments show that hydrogen transfer is direct, comparison of kinetic isotope effects with isotope discrimination studies (23) suggest that the transition state of this reaction may not be a simple transfer of hydride ion (see Section IV—2).

Shinkai and *Bruice* (21, 22) recently have described the first example of a zinc-ion-catalyzed reduction of aldehyde by NADH and NADH analogs in aqueous solution. They found that 3-hydroxypyridine-4-carboxaldehyde derivatives are reduced by 1,4-dihydropyridines in aqueous methanol (52% by weight) at 30°. Furthermore, this reaction is subject to catalysis by divalent metal ions, including Zn(II), Eq. (9). The following apparent relative order for metal ion effectivenes was observed: $Ni^{2+} > Co^{2+} > Zn^{2+} > Mn^{2+} > Mg^{2+} >$ control.

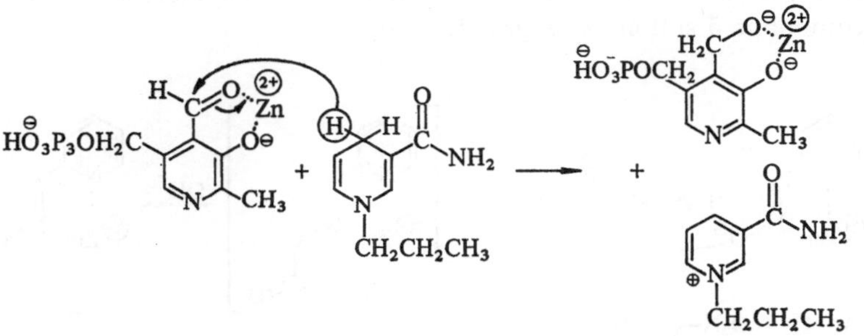

By analogy to the metal-ion-catalyzed reaction of pyridoxal derivatives with nucleophiles, *e.g.*, imine formation (24), it is highly likely that the reduction proceeds *via* the 3-hydroxypyridine-4-carboxaldehyde-metal ion complex. Coordination of the carbonyl oxygen both in the ground state and in the transition state is envisaged to facilitate hydride transfer by decreasing the electron density on the carbonyl carbon.

Thus, in the above three examples, divalent metal ion appears to lower the activation energy of the reaction both by activation of the ground-state and by stabilization of the transition-state.

2. Enhancement of Nucleophilicity, Facilitation of Nucleophilic Attack by Charge Neutralization, and the Template Effect

The investigations carried out by *Breslow* and *Chipman* (25), *Lloyd* and *Cooperman* (26), *Sigman* and *Jorgensen* (27), and *Buckingham et al.* (28), which are described below, demonstrate that divalent metal ion coordination of a weak acid can greatly increase the effective nucleophilicity of the corresponding (anionic) conjugate base. This effect is augmented when both the nucleophile and the reacting substrate are prealigned by coordination to the metal ion. In certain instances where the substrate is highly negatively charged, coordination to zinc also can assist the attack of an anionic nucleophile, since Coulombic interaction between zinc ion and substrate can effectively neutralize the (repulsive) negative charge on the substrate.

The enhancement of ligand nucleophilicity has its origins in the large pK_a' perturbations (3—4 pK_a' units) which result from coordination to the metal ion. These large pK_a' perturbations arise from the resultant Coulombic force field between the dipositively-charged zinc ion and the weak-acid dipole. This effect can be quantitatively described by Coulomb's law when the effective dielectric constant of the microenvironment is considered [*Kirkwood* and *Westheimer* (29)]. The pK_a' perturbation obviously can bring about a large increase in the effective concentration of the nucleophilic species provided the pK_a' perturbation does not decrease nucleophilicity in a manner which is proportional to the decrease in basicity.

Thus, *Breslow* and *Chipman* (25) have reported that the zinc ion-pyridine-2-carboxaldoxime anion [Zn(II)—PCA] complex is an effective and specific nucleophilic catalyst in the hydrolysis of both p-nitrophenylacetate (NPA) and 8-acetoxyquinoline-5-sulfonate (AQS), Eq. (10).

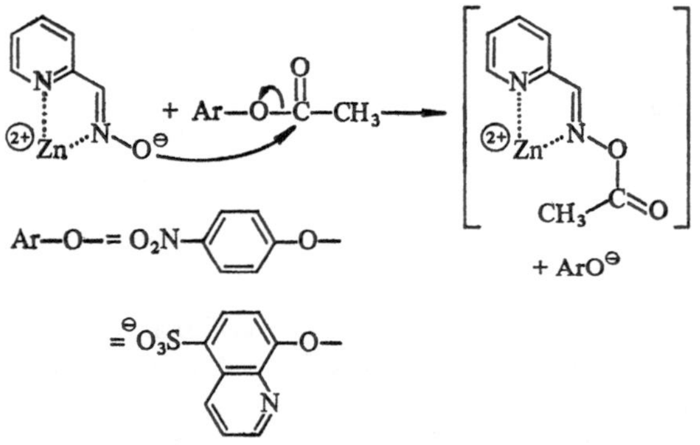

$$ \tag{10} $$

The overall rate of AQS hydrolysis in the presence of 0.01 M Zn^{2+} and 0.015 M PCA is approximately 10-fold greater than the uncatalyzed rate. Under these conditions, hydrolysis of the acetylated intermediate is ten times slower than the acetylation step. Comparison of the rates of acetylation of PCA by NPA and AQS in the presence and in the absence of zinc ion indicates that acetylation by AQS proceeds *via* the mixed zinc complex. Coordination to zinc ion perturbs the pK_a' of PCA from 10.04 to 6.5, although a nucleophilicity comparable to hydroxide is retained.

Lloyd and *Cooperman* (26) have further investigated the Zn(II)-pyridine-2-carbaldoxime anion as a catalyst for the hydrolysis of phosphorylimidazole and N-methyl phosphorylimidazole. Reaction proceeds with both compounds *via* rapid phosphoryl transfer from the phosphorylimidazole to the oximate oxygen of the Zn(II)—PCA complex. The transfer appears to occur *via* a pre-equilibrium step involving ternary complex formation, since under the conditions [Zn(II)—PCA] >> [phosphorylimidazole], the pseudo first-order rate of phosphoryl group transfer saturates. Since rate saturation occurs with virtually the same concentration dependence at both pH 4.8 and at 6.4, the imidazole nitrogen (the N-protonated mono-anion has a pK'_a of 6.8) can not be involved in complex formation. This conclusion is further supported by the observation that N-methylphosphorylimidazole is an equally good substrate. These facts suggest that both phosphorylimidazole and N-methyl-phosphorylimidazolium are coordinated to Zn(II)—PCA through the phosphoryl dianion moiety in the reactive ternary complex as illustrated in Eq. (11).

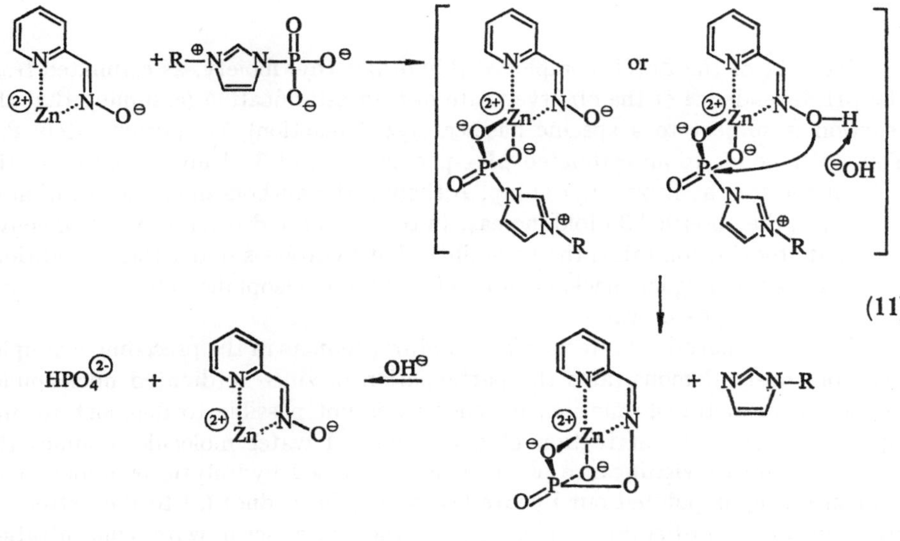

(11)

The transesterification of N-(β-hydroxyethyl)ethylenediamine by p-nitrophenyl picolinate has been shown to be subject to zinc ion catalysis by *Sigman* and *Jorgensen* (27). Their investigations indicate that reaction very probably occurs through the formation of a ternary complex in which zinc ion functions both to lower the pK'_a of the hydroxyethyl moiety, and to serve as a template for the reaction. The high specificity manifest in this catalytic process is emphasized by the fact that no catalysis of acyl-group transfer occurs when N-(β-hydroxy-ethyl)ethylenediamine is replaced by ethylenediamine, 1,5-diaminopentane, diethylenetriamine or aminoethanol. Furthermore, the reactions of the p-nitrophenyl esters of isonicotinic and acetic acids with N-(β-hydroxyethyl)ethylenediamine are not subject to zinc ion catalysis.

Zinc ion enhances the rate of transesterification relative to aqueous hydrolysis by ~180-fold, whereas hydrolysis catalyzed by zinc ion alone is increased ~33-fold, and transesterification in the absence of zinc ion is increased ~11-fold over hydrolysis. While pointing out that these data do not uniquely define the structure of the reactive complex, *Sigman* and *Jørgensen* suggest that reaction proceeds *via* the ternary complex shown below:

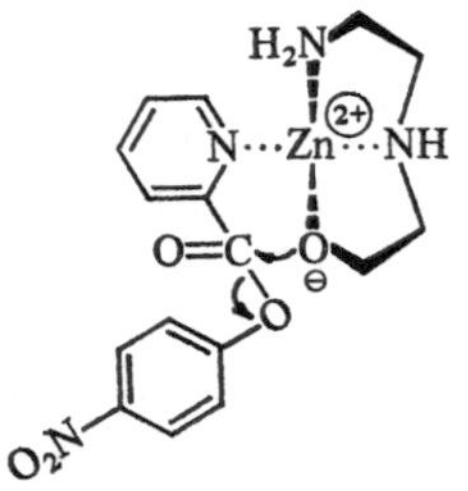

Reactive ternary complex

The pK_a' of the Zn(II)-complexed β-hydroxyethyl moiety, as estimated from the pH-dependence of the observed rate of transesterification (assuming that the reaction is subject to a specific base-catalyzed reaction), is approximately 8.4. This corresponds to an estimated pK_a' perturbation of 3—4 units relative to the pK_a' of the free hydroxyethyl group. Although the authors make no mention of this point, the observed 33-fold increase in the rate of hydrolysis of p-nitrophenyl-picolinate by zinc ion raises the possibility that hydrolysis under these conditions occurs *via* an analogous mechanism involving the nucleophilic activation of coordinated (inner sphere) water.

While the analysis of intermediates and/or products in the preceding examples unambiguously demonstrates the participation of zinc-coordinated nucleophiles in the chemical transformation, it generally is not possible to demonstrate unequivocally that the activation of a coordinated water molecule assumes the same mechanistic significance in metal ion-catalyzed hydrolytic reactions. This happenstance, as pointed out by *Breslow et al.* (*19*), is due: (a) to the extremely labile nature of most aquo complexes, (b) to the invariance of water concentration in aqueous milieu, and (c) to the mechanistic ambiguity arising from the kinetic equivalance of the two mechanisms, *A* and *B*, illustrated below (here p-nitrophenyl picolinate is used as the hypothetical example).

Buckingham et al. (*28*) have been able to unambiguously show that the cobalt-(II)-catalyzed hydrolysis of glycine esters proceeds in dual pathways at ap-

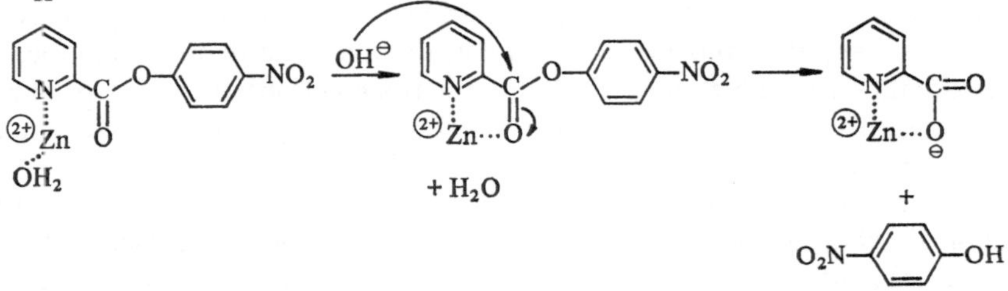

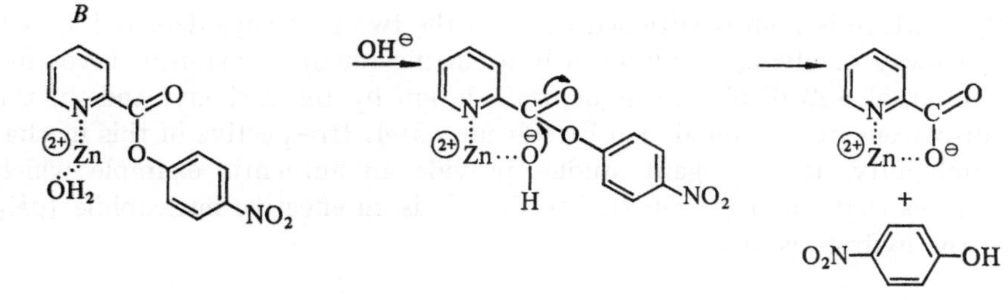

$$(13)$$

proximately equal rates by mechanisms corresponding to A and B. In their investigation of the $[Co(en)_2X(Gly-OR)]^{2+}$ system, they find that only the halide ligand (X) is replaced by other ligands in aqueous solution during the course of the hydrolysis of Gly-O-Et to bound Gly and free ethanol. In the alkaline pH range, hydrolysis yields $[Co(en)_2Gly]^{2+}$ in a two-step process. Both steps are characterized by kinetic dependencies which are first-order in complex and first-order in hydroxide ion. Kinetic and product studies indicate that these two steps correspond to substitution of halide by H_2O and to ester hydrolysis. Since these Co(III) complexes are exchange inert (that is, ligand exchange is a slow process) ^{18}O-labeling experiments were used to distinguish between mechanisms A and B. Thus, when the reaction was run in ^{18}O-enriched water, the product was found to have incorporated close to one oxygen atom from the solvent. However, when carbonyl-^{18}O-$[Co(en)_2GlyOCH(CH_3)_2]^{3+}$ is hydrolyzed in normal water, 45% of the ^{18}O label is retained in the product. Furthermore, neither carbonyl-^{18}O-$[Co(en)_2GlyOCH(CH_3)_2]^{3+}$ nor carbonyl-^{18}O-$[Co(en)_2Gly]^{2+}$ exchange ^{18}O with the solvent and the position of the label does not scramble during the time required for hydrolysis. *Buckingham et al.* (*28*) conclude that the $\sim$50% distribution of label must arise from competing pathways which contribute equally to product formation. These pathways would appear to correspond to the two reactions illustrated below:

As the authors point out, the available data do not distinguish between a pathway involving a common penta-coordinate intermediate, *i. e.* $[Co(en)_2$-

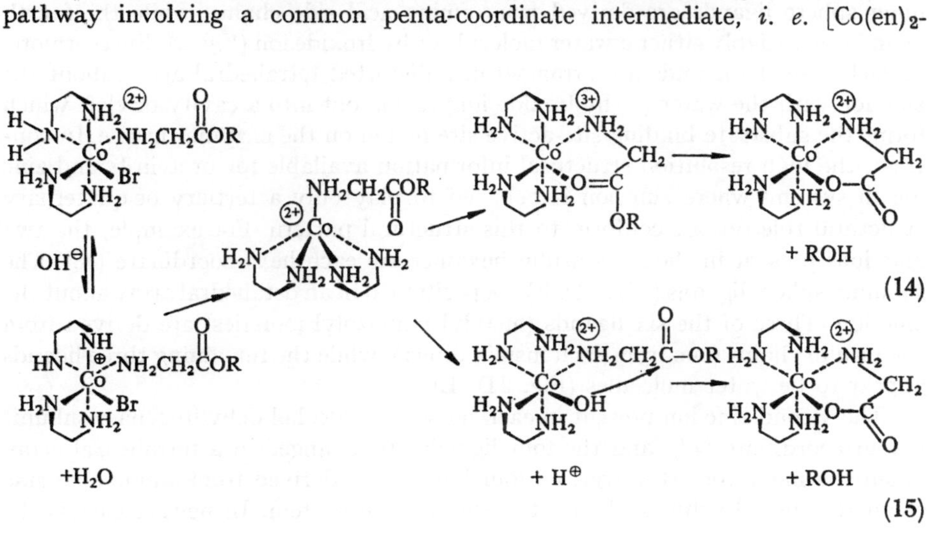

$$(14)$$

$$(15)$$

GlyOR]$^{3+}$, which is then partitioned between the two pathways depicted above, and a pathway involving simultaneously occurring (dual) mechanisms involving (a) the internal S_N2 displacement of bromide ion by the carbonyl and (b) the S_N1B displacement of bromide ion by solvent water. Irrespective of this mechanistic ambiguity, these elegant studies provide an authentic example which demonstrates that water coordinated to Co(III) is an effective nucleophile (pK_a' ~6) in the hydrolysis of glycine esters.

IV. Catalysis by Zinc Metalloenzymes

The general physical and chemical properties of several of the zinc-metalloenzymes discussed here have been the subjects of excellent comprehensive reviews by *Coleman (30), Lindskog et al. (31), Lindskog (32), Hartsuck* and *Lipscomb (33), Quiocho* and *Lipscomb (34)*, and *Sund* and *Theorell (35)*. Therefore, it would serve no purpose to attempt a comprehensive review of zinc-metalloenzymes. For this reason, the present work is confined primarily to a review of the literature which pertains to the role(s) played by zinc ion in enzyme catalytic mechanism. This work is further restricted by limiting discussion to consideration of only those enzyme systems for which high resolution 3-dimensional X-ray structures are available.

1. Structural and Chemical Properties of Zn-Metalloenzymes

The high resolution, three-dimensional structures for the active sites of carboxypeptidase *(33, 34, 36—39)*, carbonic anhydrase *(31, 40—42)*, thermolysin *(43—46)*, and horse liver alcohol dehydrogenase *(47)* exhibit several structural features of the zinc(II) coordination sphere which appear to be uniquely characteristic for catalytic zinc ion. In each of the above mentioned structures, the active site zinc ion is four-coordinate in the native enzyme, as shown in Fig. 2. Three of the four inner sphere ligands are derived from amino acid side chains, while the fourth ligand is invariably either a water molecule or hydroxide ion (Fig. 2). Furthermore, in each case the ligands are arranged in a distorted tetrahedral array about the zinc ion with the water (or hydroxide ion) facing out into a cavity or cleft which forms the substrate binding site-active site region on the enzyme surface. In contrast, the high resolution structural information available for protein-bound zinc ion in systems where zinc ion is believed to play only a tertiary or quaternary structural role do not conform to this structural pattern. For example, the two zinc ions present in the zinc-insulin hexamer are each hexa-coordinate *(48)*. The six inner sphere ligands (Fig. 2 D, E) are positioned in an octahedral array about the zinc ion. Three of the six ligands (histidyl imidazolyl moieties) are derived from the insulin chains (one from each insulin dimer), while the remaining three ligands appear to be water molecules (Fig. 2 D, E).

The second zinc ion present in each horse liver alcohol dehydrogenase subunit is four-coordinate *(47)*, and the four ligands are arranged in a tetrahedral array about the metal ion. However, all four ligands are derived from amino acid side chains (cysteinyl sulfhydryls) contributed by the protein. In neither case is the

M. F. Dunn

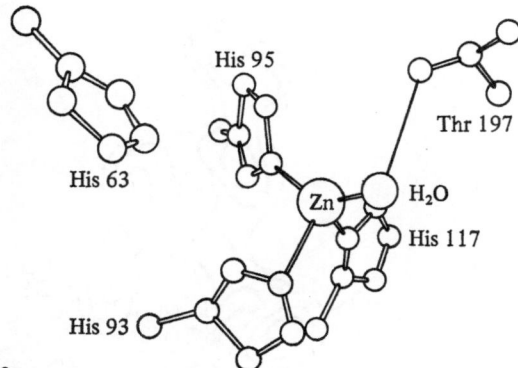

Fig. 2 A. Carbonic anhydrase

Fig. 2 B. Carboxypeptidase

Fig. 2 C. Thermolysin

Fig. 2. Three-dimensional structural representations for zinc metall-oproteins. Comparison of the zinc ion-protein bonding interactions for zinc requiring enzymes (A—C) with the zinc-insulin hexamer (D, E). (A) Human carbonic anhydrase C, redrawn from Ref. (41) with permission. (B) Bovine carboxypeptidase Aγ, redrawn from Ref. (30) with permission. (C) *Bacillus thermoprotedyticus* thermolysin, redrawn from Ref. (45) with permission. (D) and (E) Porcine Zn-insulin hexamer, taken from Ref. (48) with permission. The composite electron density maps in (D) and (E) show that each of the two zinc atoms present in the hexamer is within inner sphere bonding distance of three solvent molecules and three histidyl imidazolyl groups in an octahedral array about the metal ion. The position of one of the three equivalently positioned solvent molecules is indicated in (D). The electron density map in (E) shows the relative orientations of the three histidyl residues (His-B10). (The atomic positions of one of the three equivalent histidyl groups are shown)

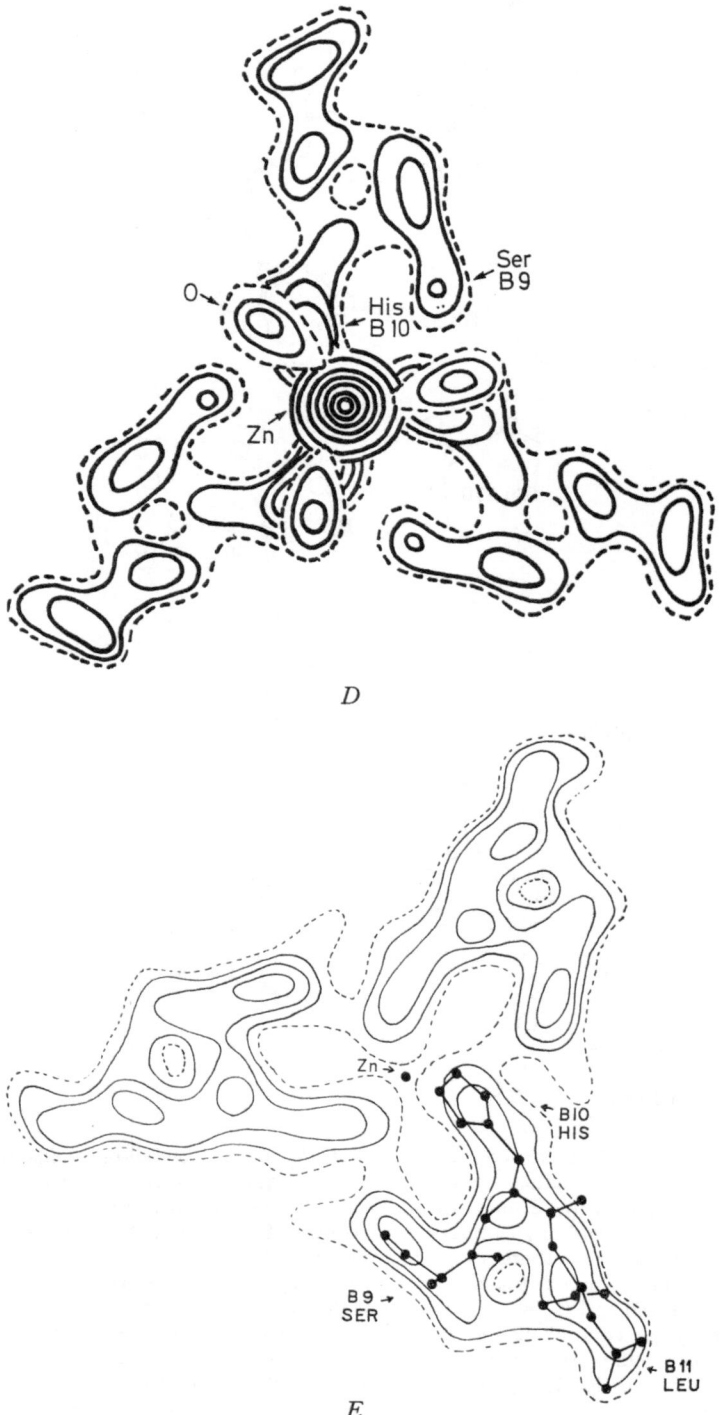

Fig. 2 D, E. Insulin

zinc ion located at or near an obvious active site and/or ligand binding site cavity or cleft in the protein. It should be noted in this context that the two zinc ions present in the insulin hexamer lie on the three-fold symmetry axis, and that they are separated by an internal cavity which forms a core about which the protomers are situated. This core is occupied by a highly structured solvent matrix which is presumed to consist of water molecules arranged in a hydrogen-bonded (ice-like) lattice (48).

The amino acid side chain residues which have been identified *via* X-ray structure determinations as ligands for active site zinc ion (see Fig. 2) include the histidyl imidazolyl moiety (*e.g.*, carboxypeptidase, carbonic anhydrase, horse liver alcohol dehydrogenase), the cysteinyl sulfhydryl group (horse liver alcohol dehydrogenase), and the glutamyl carboxylate moiety (carboxypeptidase and thermolysin). It is not surprising that these amino acid side chains are involved in zinc-metalloenzyme coordination complexes since the donor atoms involved, amino nitrogen, carboxylate oxygen, and thiolate sulfur, form relatively affine complexes with zinc ion in aqueous solution. Within the relatively large limits of error inherent in the 3-dimensional protein X-ray structure determinations, the donor-atom zinc-ion bond-lengths in each case are reported to be normal: 2.0—2.2 Å for N-Zn, and 2.0—2.1 Å for O-Zn bonds, (compare with Fig. 1).

The presence of an inner sphere water molecule (or hydroxide ion) at the fourth ligand site for catalytically essential zinc ion introduces questions which are central to an understanding of the role(s) played by zinc ion in the catalytic mechanism. Does zinc ion activate the water molecule for direct participation in the chemical transformation — *i.e.*, as an acid-base catalyst, or as a nucleophile?; or, does zinc ion function as a Lewis acid in the activation of substrate chemical bonds? In the latter case, inner sphere coordination of substrate most probably would be preceded by displacement of the inner sphere water molecule, although ligand exchange mechanisms involving a penta-coordinate transition-state and/or penta-coordinate intermediate(s) can not be excluded *a priori* as possibilities.

Neither the X-ray structural information nor the kinetic information in hand unambiguously resolves this question for any one of the enzymes for which high resolution structural information is available. This is partially a consequence of the ambiguity inherent in any attempt to construct a catalytic mechanism based primarily on a structural and/or chemical knowledge of stable (ground-state) enzyme-inhibitor or enzyme-quasisubstrate complexes. Note that these considerations are equally apropos to other physico-chemical techniques (*e.g.*, UV-visible, IR, NMR, and ESR spectroscopic techniques) when applied to the investigation of those species detectable at equilibrium. Furthermore, the available kinetic information pertaining to the composition of the transition state(s) for the chemical step(s) is insufficient to allow the resolution of this question.

It is important in the context of this discussion to note that those inhibitors or quasisubstrates which are both good structural and good chemical analogs of substrate are almost without exception weak ligands for zinc (as invariably is the case for substrate). Consequently, intermediates on the reaction pathway to products which conceivably could involve inner sphere coordination between substrate (or analog) and the active site zinc ion may not be easily detected by direct observation since they are likely to be relatively high energy species.

It is primarily for the above reasons that, in the view of this author, it is not yet possible to unequivocably define the mechanistic role played by zinc ion for any zinc-enzyme. Nevertheless, with the exception of thermolysin, it is possible to arrive at reasonable mechanistic hypotheses for the various zinc enzymes considered in this review through the examination of data derived from both kinetic studies and from studies at equilibrium through judicious application of the "anthropomorphic" approach to the description of reaction mechanisms (50).

2. Alcohol Dehydrogenase from Equine Liver

The alcohol dehydrogenase isolated from horse liver (LADH) [E.C.1.1.1.1.], MW 80,000, is available as a highly purified, crystalline material composed of two (or more) isozymes (51), one of which is present in >90%. The general physical and chemical properties of this zinc enzyme have been reviewed by *Sund* and *Theorell* (35). This nicotinamide-adenine dinucleotide oxidoreductase catalyzes the aldehyde-alcohol interconversion shown in Eq. (16).

$$RCHO + NADH + H^{\oplus} \underset{\overrightarrow{}}{\overset{K_{eq}}{\rightleftharpoons}} RCH_2OH + NAD^{\oplus} \qquad (16)$$

$$K_{eq} = \frac{[RCH_2OH][NAD^{\oplus}]}{[RCHO][NADH][H^+]}$$

Equilibrium is thermodynamically in favor of alcohol formation in the physiological pH range for most aliphatic and aromatic aldehydes [*i.e.*, $K_{eq} = 1.25 \times 10^{12} M^{-1}$ for acetaldehyde (35)]. While it has often been assumed that ethanol and acetaldehyde are the physiologically significant substrates, there appears to be as much evidence against the validity of this assumption as there is for it. Alternative suggestions for the physiological role of this enzyme range from the production of steroid alcohols (51) to the oxidation of ω-hydroxy fatty acids (52) to a general detoxification role in the liver.

The specificity of the horse liver enzyme for substrate is rather broad: turnover numbers for aldehydes of widely varying structure (*e.g.*, acetaldehyde *vs.* β-napthaldehyde) exhibit only a 25-fold variation in k_{cat}. In contrast, coenzyme specificity is quite restrictive. The early studies of *Anderson et al.* (53, 54) and more recently of *Biellmann et al.* (55) indicate that the nicotinamide ring can be chemically altered only at the 3-position without destroying the chemical functionality of the dinucleotide. Furthermore, the activity of coenzyme analogs chemically altered at the 3-position is highly sensitive to the electronic and steric nature of the chemically modified substituent (55).

The native (dimeric) enzyme consists of two (40,000 dalton) subunits composed of covalently identical amino acid chains (47, 56). Each subunit has been shown to contain two gram-atoms of zinc ion per subunit, one of which is essential for catalytic activity (57—64). And each subunit, as will be discussed in more detail below, contains a coenzyme binding site which is adjacent to one of the two zinc ions (47).

The pH dependence of the apparent dissociation constants for the binary enzyme-NAD+ and enzyme-NADH complexes for the native enzyme are compared

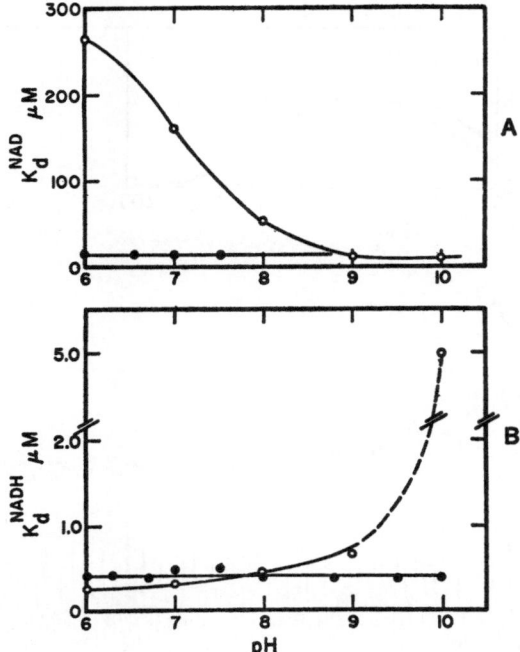

Fig. 3. Dissociation constants of NAD (A) and NADH (B) from apo- (●) and native (○) enzymes as a function of pH. The K_a's for NAD from apoenzyme were obtained by equilibrium dialysis and for NADH by fluorescence polarization. The K_d's for NAD and NADH from the native enzyme were obtained from Ref. (65) with permission. The figure is taken from Ref. (63) with permission

in Fig. 3 to the corresponding apparent dissociation constants for the zinc-free (apo-) enzyme (63, 65). The remarkable differences in the pH dependence of the affinities of native enzyme for oxidized and reduced coenzymes are strongly indicative of the interaction of one (or more) ionizable groups at the site with bound coenzyme. *Taniguchi et al.* (66) have calculated that the pH-dependencies of coenzyme binding to the native enzyme can be interpreted as resulting from the perturbation of a single residue at the site of $pK_a' \sim 8.6$ in the native enzyme. *Dalziel* (67) has presented evidence from steady-state kinetic studies which indicate that there may be more than one ionizable residue involved. The lowering of the intrinsic pK_a' of one (or more) ionizable groups by as much as 1.9 pK_a' units in the enzyme-NAD$^+$ complex (*viz.* Fig. 3) must almost certainly arise from Coulombic interactions between the positively-charged nicotinamide moiety of NAD$^+$ and the ionizable residue(s) when placed in close proximity.

The existence of such a pK_a' perturbation finds additional support from kinetic studies (Fig. 4) which unequivocably establish that the displacement of enzyme-bound NAD$^+$ by NADH at pH 8.8 is accompanied by the net uptake of approximately one mole of hydrogen ion from solution per mole of NAD$^+$ displaced (68).

The increase in the apparent pK_a' of the ionizable residue from $\sim$8.6 in the native enzyme to $\sim$9.5 in the enzyme-NADH complex (assuming a single ionizable group is involved) can be attributed to a screening effect. The neutral, essen-

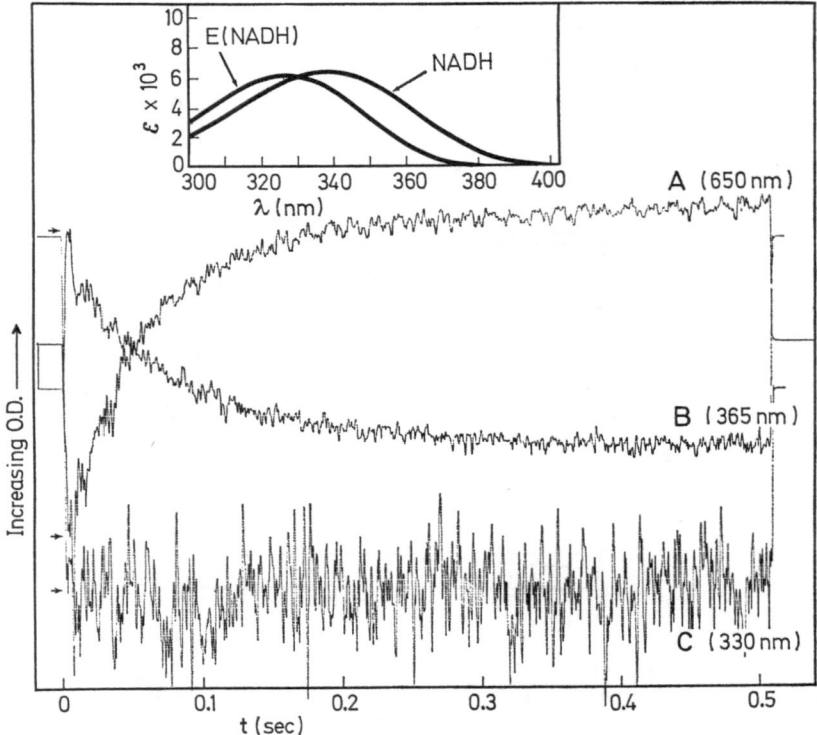

Fig. 4. Representative stopped-flow traces comparing the time course for the formation of the enzyme-NADH binary complex measured at 365 nm (trace A) with the time course for the uptake of hydrogen ions as measured by the Thymol Blue spectral changes at 650 nm (trace B). Trace C measures the optical density at 330 nm, the isobestic point for the two NADH species: enzyme bound and free in solution, see insert to this figure. Traces A and C were measured in 0.5 mM sodium pyrophosphate buffer (pH 8.80 ± 0.1). In trace B, 0.65 mM Thymol Blue, pH 8.8 ± 0.1, is the dominate buffer ion present in solution. Conditions: E, 10.6 μN; NAD$^+$, 23.9 μM; NADH, 94.1 μM; Thymol Blue, 0.384 mM; pH 8.8 ± 0.1 and 25.0 ± 0.2°. The net optical density changes (corrected to the values expected for a 1-cm light path) at 365 and 650 nm are 0.0120 and 0.0125 optical density, respectively. The insert to this figure compares the spectrum of NADH free in solution with the spectrum of enzyme-bound NADH. Taken from Ref. (68) with permission

tially non-polar nicotinamide moiety of NADH is envisaged to effectively screen the ionizable group from the aqueous/polar environment of the coenzyme-free site, thereby increasing the free energy requirements for ionization within the complex.

The recent work of *Iweibo* and *Weiner* (62) and *Coleman et al.* (63) provide evidence which shows that zinc ion plays an essential role in determining the pH dependencies of coenzyme binding. Their work (see Fig. 3) demonstrates that the aforementioned pH-dependence of site affinity for coenzyme is lost in the zinc-free (apo-) enzyme. However, note in Fig. 3 that the apo-enzyme retains the same high affinities both for NAD$^+$ and for NADH exhibited by the native enzyme at the respective pH optima for complex formation. Thus, while the pH-dependence

81

of coenzyme binding is critically dependent on the presence of zinc ion at the active site, it is unlikely that coenzyme binding involves inner sphere coordination to zinc ion. As will become apparent in the discussion which follows, the features of the three-dimensional structure at 2.4 Å resolution (47) leave little doubt but that the coordinated water molecule is the ionizable group which regulates the pH dependence of NAD+ binding.

With the exception of the recent equilibrium binding studies of *Everse* (69) there is no evidence from equilibrium binding studies for the presence of either homotropic or heterotropic allosteric binding interactions between coenzyme and/or substrate binding sites on the two subunits. Nevertheless, the presence of bound coenzyme has a profoundly "cooperative" effect on the affinity of the enzyme for substrates or substrate-analogs which compete for the substrate binding site. The synergism between coenzyme binding and substrate (or analog) binding *at the same site* is manifest in large increases in the apparent affinity of the enzyme site for both coenzyme and substrate, or analog (35, 65, 70—72). Furthermore, the ligand binding specificity of the coenzyme-enzyme complex is critically dependent upon the coenzyme oxidation state. These oxidation-state-dependent differences in ligand affinities adhere to a pattern which appears to reflect the chemical differences between bound NAD+ and bound NADH. For example, aliphatic and aromatic primary amides form complexes only with the binary E-NADH complex, while aliphatic carboxylic acid anions exhibit a high affinity only for the binary E-NAD+ complex (35, 70). In the latter case, a highly favorable Coulombic interaction within the ternary complex involving the carboxylate anion and the pyridinium ion moiety of NAD+ no doubt accounts for the large affinity increase. While the factors which contribute to the stabilization of the enzyme-NADH-amide ternary complexes are not as obvious, there is strong circumstantial evidence in support of the view that amide affinity is primarily a function of the lipophilic character of the amide, and secondarily, a function of the polarizability of the amide (73, 74). Since electron-releasing substituents generally contribute to an increased affinity for benzamides, *Sarma* and *Woronick* (74) and *Hansch et al.* (73) conclude that the amide carbonyl moiety interacts with the site through the amide carbonyl oxygen (presumably *via* inner sphere coordination to the active site zinc ion).

Comparison of equilibrium constants for the dissociation of acetaldehyde from the enzyme site in the absence of coenzyme ($K_s \simeq 2.7 \times 10^{-1}$M) (75) with the Michaelis constant for acetaldehyde ($K_m^s = 2.1 \times 10^{-4}$M (35)) indicates that the presence of bound coenzyme increases the affinity of the site for this substrate $\geq$50-fold.[1]) Somewhat larger differences in apparent affinity are found for other substrates (35, 72, 75, 78).

---

[1]) The Michaelis constant for substrate measures the overall affinity of the E(coenzyme) binary complex for all substrate species on the reaction pathway. This gross "affinity" is expressed by the relationship:

$$K_m^s = \frac{[E(coenzyme)] \, [S]}{[E_0] - [E]} = \frac{[E(coenzyme)] \, [S]}{[E(coenzyme, S)] + [E(coenzyme, P)] + \sum\limits_{i=1}^{n} [EX_i]}$$

where P = product and
X_i = the ith intermediate on the reaction pathway.

In the case of the isobutyramide-NADH-enzyme ternary complex, the affinity of the enzyme-NADH binary complex for isobutyramide is increased by a factor of $\sim 10^2$ relative to that of the native enzyme in the absence of NADH (*35, 70, 75*), and the apparent affinity of the enzyme for NADH is increased by a similar factor in the presence of saturating isobutyramide concentrations (*35, 70*).

As demonstrated by the classical steady-state kinetic studies of *Theorell* and *Chance* (*76*), and by the elegant studies of *Wratten* and *Cleland* (*77*) and *Dalziel* (*67*), the synergism between coenzyme and substrate binding conveys a kinetically discernible sequence to the binding of coenzyme and substrate. The binding of coenzyme is an obligatory (compulsory) step for substrate binding. Release of products from the enzyme site occurs *via* the microscopic reversal of this sequence. The sequential relationship between binding steps and the redox step(s) for the overall transformation is summarized in Eqs. (17–21).

$$E + NADH \rightleftharpoons E(NADH) \tag{17}$$

$$E(NADH) + S \rightleftharpoons E(NADH,S) \tag{18}$$

$$H^\oplus + E(NADH,S) \rightleftharpoons \ldots \rightleftharpoons E(NAD^\oplus,P) \tag{19}$$

$$E(NAD^\oplus,P \rightleftharpoons E(NAD^\oplus) + P \tag{20}$$

$$E(NAD^\oplus) \rightleftharpoons E + NAD^\oplus \tag{21}$$

In the physiological pH range, the specific first-order rates of dissociation of both NAD^+ and NADH from their respective binary complexes are slow relative to the rate of the chemical transformation. Indeed, *Bernhard et al.* (*78*) have shown that for aldehyde reduction the chemical step occurs at an overall rate which is one to two orders of magnitude greater than the steady-state turnover rate. Hence, for many substrates, the velocity of substrate turnover under steady-state conditions is limited by the rate of dissociation of coenzyme product (*78–80*). The steady-state kinetic studies of *Wratten* and *Cleland* (*77*) and the rapid (transient-state) kinetic studies of *Bernhard et al.* (*78*), *McFarland* and *Bernhard* (*80*), *Luisi* and *Favilla* (*81*), and *Dunn* (*68*) have shown that for certain aromatic aldehydes, release of the aryl alcohol product from the site determines the steady-state rate.

In recent years, the stopped-flow rapid-mixing kinetic technique and, to a lesser extent, the temperature perturbation of equilibrium (temperature-jump) technique have been used by a number of workers (*68, 72, 79–86*) to investigate individual steps in the above reaction scheme. Much of this work has been directed toward the investigation of the relationship between catalytic mechanism and subunit function in the dimeric enzyme. Since the scope of this review is limited to consideration of the involvement of zinc ion in the mechanism of the chemical transformation, no discussion of subunit function is presented here. Those readers who wish to pursue this aspect of LADH catalysis are referred to the original literature (*68, 72, 79–86*) and to the excellent reviews which have recently appeared on this subject (*87, 88*).

The rapid kinetic investigations of *Bernhard et al.* (*78*) were carried out under experimental conditions which limit the extent of reaction to a single turnover of those sites which encounter substrate by placing enzyme and one reactant

(NADH or aldehyde) in excess of the other reactant. Under these conditions, the chemical conversion of aldehyde to alcohol occurs with a (saturated) apparent first-order rate constant of 200 to 400 sec^{-1}. This process, as measured either by the disappearance of NADH or by the disappearance of chromophoric aldehyde, has been shown by *McFarland* and *Bernhard* (*80*) to be subject to a primary, kinetic isotope effect $k_H/k_D = 2$ to 3 when stereospecifically labeled (4-R)-deuterio NADH is compared to isotopically normal NADH. *Shore* and *Gutfreund* (*84*) earlier had investigated substrate kinetic isotope effects on the pre-steady-state phase of ethanol oxidation. Their studies demonstrated that the rate of the pre-steady-state burst production of NADH is subject to a primary kinetic isotope effect, $k_H/k_D \simeq 4$—6 when 1,1-dideuterio ethanol is compared to isotopically normal ethanol, and that there is no primary kinetic isotope effect on the steady-state rate. It can be concluded from these studies: (a) that the rate of interconversion of ternary complexes [*e.g.*, Eq. (19) above], as already mentioned, is rapid relative to turnover, and (b) that the transition-state for the rate-limiting step in the interconversion of ternary complexes involves carbon-hydrogen bond scission and/or carbon-hydrogen bond formation.

Jacobs et al. (*86*) have compared the electronic substituent effect on the rate of sodium borohydride reduction *vis à vis* the LADH-catalyzed reduction (under transient-state kinetic conditions) for a series of para-substituted benzaldehydes. In contrast to the large electronic substituent effect observed in the sodium borohydride reaction (the rate ratio k_{p-Cl}/k_{p-OCH_3} is ~100), the LADH catalyzed reaction was found to show almost no electronic substituent effect ($k_{p-Cl}/k_{p-OCH_3} \simeq 2$). As these authors point out, the near absence of an electronic substituent effect is unexpected in view of the above-mentioned primary kinetic isotope effects if reaction is assumed to be a simple hydride transfer process. These authors (*86*) suggest that the unexpectedly low substituent effect may reflect a transition-state for the enzymatic process which, in comparison to borohydride reduction is considerably shifted along the reaction coordinate toward the E(NADH, ald) ternary complex. Assuming a hydride-transfer process obtains, these authors point out that any interaction with the carbonyl group which significantly increases the positive character of the carbonyl carbon could explain the low electronic substituent effect. On the basis of this argument, they conclude that a mechanism in which zinc ion acts as a Lewis acid catalyst in facilitating hydride transfer between coenzyme and substrate is consistent with the small substituent effect. However, it should be noted that the small electronic substituent effect also is consistent with a free radical mechanism involving hydrogen atom transfer in the transition-state of the reaction.

The stopped-flow rapid-mixing and temperature-jump kinetic studies from the author's laboratory (*72, 89*) which describe the reaction of the enzyme-NADH complex with the intense chromophore, *trans*-4-N,N-dimethylaminocinnamaldehyde (DACA) (λ_{max} 398 nm, ε_{max} 3.15 $\times$ 10^{4}M^{-1}cm^{-1}) provide direct evidence for the involvement of zinc ion as a Lewis acid in the activation of the aldehyde carbonyl for reduction. These studies show that reduction of DACA involves the formation of a transient chemical intermediate (λ_{max} 464 nm, ε_{max} ~6.2 $\times$ 10^{-4}M^{-1}cm^{-1}) in the neutral pH range. At pH values above 9 and in the presence of high enzyme concentrations, the transient species observed

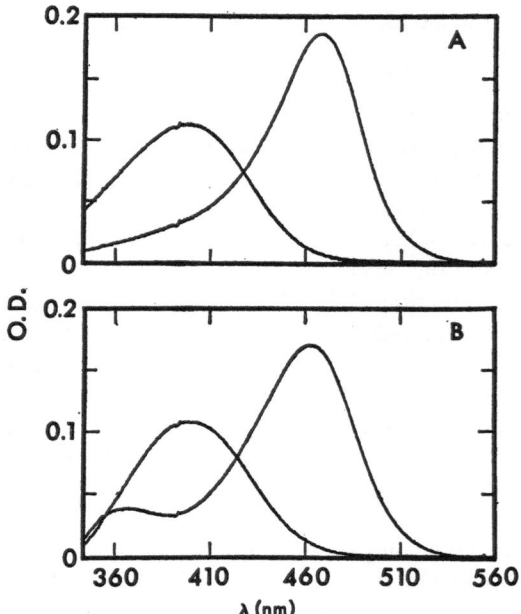

Fig. 5. UV-visible spectral properties of the DACA-enzyme-H$_2$NADH and DACA-enzyme-NADH systems. The traces in (A) compare the spectrum of DACA (λ_{max} 398 nm) with the spectrum of the intermediate (λ_{max} 468 nm) derived from the reaction of DACA with the enzyme-H$_2$NADH complex on 0.1 M sodium phosphate buffer, pH 6.98, at $25 \pm 1°$. These spectra were measured using double difference (split compartment) cuvettes. The configuration of solutions used in the determination of the spectra in (A) are as follows: For the spectrum of DACA the sample cuvette contains enzyme, 85.5 μN and H$_2$NADH, 286 μM in one compartment (total vol. 1.10 ml) and DACA, 8.02 μM on the second compartment (total vol. 1.10 ml); the reference cuvette contains an equal volume of the identical enzyme-H$_2$NADH solution in one compartment, and 1 ml of buffer in the other compartment. The spectrum of the intermediate was obtained by mixing the contents of the sample cuvette and recording the spectrum of the resulting reaction mixture. Note that under these experimental conditions, the conversion of DACA to intermediate is nearly complete. The traces in (B) compare the spectrum of DACA with the spectrum of the intermediate (λ_{max} 464 nm) derived from the reaction of DACA with the enzyme-NADH complex in 0.1 M sodium carbonate buffer, pH 9.63, at $25 \pm 0.1°$. The spectrum of DACA and the spectrum of the intermediate were obtained as in (A). The conversion of DACA to intermediate also is nearly complete under these conditions; furthermore, the net conversion of intermediate to products is negligible at this pH

at lower pH values becomes a relatively stable species (Fig. 5). Below pH 9, the intermediate decays in a pH-dependent step to an equilibrium mixture consisting of *trans*-4-N,N-dimethylaminocinnamyl alcohol, NAD$^+$, and intermediate at a rate which is 2 to 3 orders of magnitude slower than the apparent rate of intermediate formation [Eq. (22, 23)]. Investigation of the mechanism of intermediate formation *via* uv-visible spectroscopy and temperature-jump kinetic studies (72) show that intermediate formation is a simple, reversible process [Eq. (22)] which is pH-independent over the pH range 6 to 10.5.

$$E(NADH) + DACA \underset{k_{-1}}{\overset{k_1}{\rightleftharpoons}} E(\text{Intermedaite}) \tag{22}$$

$$E(\text{Intermediate}) + H^+ \underset{\longleftarrow}{\overset{\longrightarrow}{}}\cdots\cdots\underset{\longleftarrow}{\overset{\longrightarrow}{}} E + NAD^+ + \text{Alcohol} \tag{23}$$

At 31°, the specific rate constants for the forward and reverse process are $k_1 = 4$ to $6 \times 10^7 M^{-1}$ sec^{-1} and $k_{-1} \simeq 280$ sec^{-1}. Subsequent investigations (89) have shown that the coenzyme analog 1,4,5,6-tetrahydronicotinamide-adenine dinucleotide (H$_2$NADH) will substitute for NADH in the formation of a stable chromophoric species (λ_{max} 468 nm, ε_{max} 5.8 × 10^{4}M^{-1}cm^{-1}) (see Fig. 5) which is structurally analogous to the transient species formed in the presence of NADH. This is a particularly striking finding since H$_2$NADH is chemically inert in the LADH-catalyzed oxidoreduction reaction (90). Neither NAD$^+$, adenosine-di-phosphoribose (ADPR), nor adenosine monophosphate (AMP) will function in place of NADH (or H$_2$NADH) in the process of intermediate formation. Thus, for this system, the previously noted oxidation-state-dependent differences in ligand affinities is manifest in the form of an absolute requirement for reduced dinucleotide in order for the first step, Eq. (22), to occur.

The chemical properties of the 1,4,5,6-tetrahydronicotinamide ring virtually exclude the possibility that formation of the chromophore involves covalent chemical bond formation between DACA and reduced dinucleotide. Therefore, we have concluded that reduced dinucleotide plays a heretofore unsuspected effector role in facilitating this reaction. Both NADH and H$_2$NADH bring about the same crystallographically identifiable conformational change in LADH, (E. Zeppezauer, C.-I. Bränden and H. Ekland, unpublished results). Together, these findings directly demonstrate that coenzyme binding stabilizes the catalytic-ally active enzyme conformation state, and thereby plays a noncovalent role in effecting the chemical activation of substrate.

Dunn and Hutchinson (72) and Dunn et al. (89) conclude that only a structure involving inner sphere coordination of the carbonyl oxygen of DACA to the active site zinc ion, as depicted in Scheme I, can explain: (a) the pH-independence of the process, (b) the magnitude of the spectral red shift in the chromophone spectrum (60—70 nm), (c) the rapid formation rate constant ($\sim$4 × 10^{7}M^{-1} sec^{-1}), and (d) the absence of a primary deuterium isotope effect on the rate of interme-diate formation.

The high resolution X-ray structural studies of the native enzyme, the enzyme-ADPR binary complex, and the enzyme-o-phenanthroline binary complex (47) have revealed that the active site zinc ion is located some 20 Å below the surface of the protein at the point of convergence of two deep clefts (see the schematic representation in Fig. 6). One of these clefts has been identified as the coenzyme binding cleft (47). This cleft extends from the surface of the subunit to the zinc ion. If a model of NADH is fit to the coordinates of the ADPR binding site, then the nicotinamide ring can be oriented in such a way that it fits into a pocket adjacent to the zinc ion (47). The second deep cleft, or channel, also extends from the surface of the subunit down to the zinc ion. The inner surface of this cleft is made up of nonpolar amino acid residues contributed by both subunits.

SCHEME I

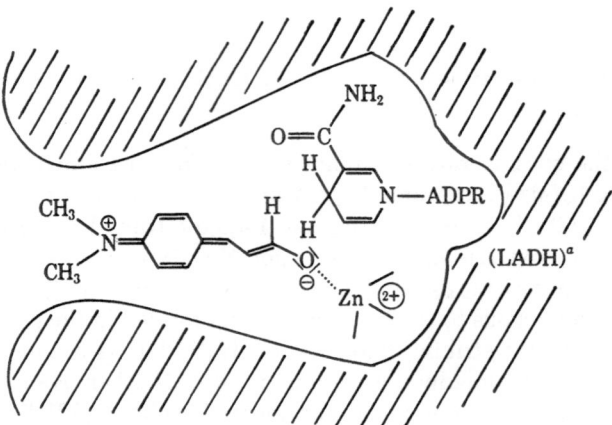

a LADH = horse liver alcohol dehydrogenase.

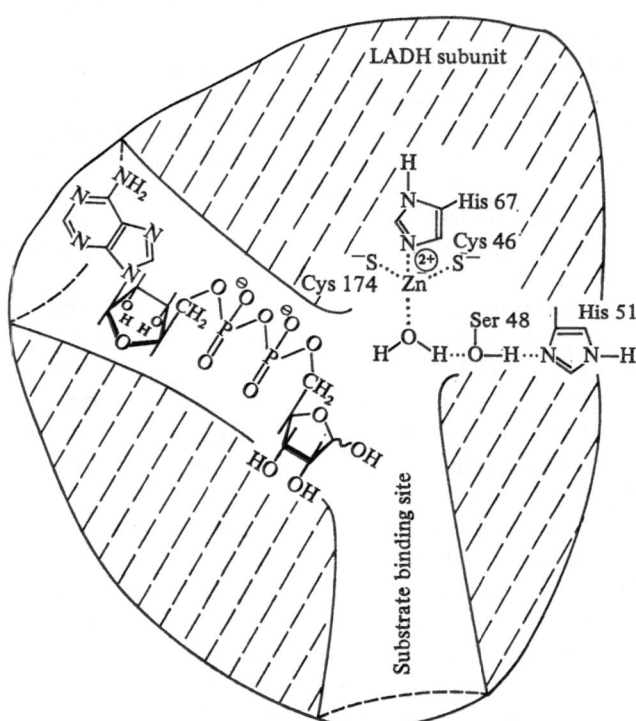

Fig. 6. Sketch of the horse liver alcohol dehydrogenase-adenosine-diphosphoribose (ADPR) binary complex in the region of the active site. ADPR lies in an ~20 Å-deep cavity which extends to the metal ion (as indicated). The second cavity (~20 Å deep also) is believed to be the substrate binding site (*47*)

The structure of the enzyme-o-phenanthroline binary complex shows that o-phenanthroline occupies this cleft while coordinated to the active site zinc ion. Note that the LADH zinc ion o-phenanthroline complex is pentacoordinate.

As already described in the general discussion which precedes this section, the active site zinc ion of the native enzyme has a water molecule coordinated to the fourth ligand site. This water molecule is positioned within hydrogen-bonding distance of the hydroxymethyl oxygen of serine residue 48. The serine hydroxyl group in turn is located within hydrogen bonding distance of the N-1 nitrogen of a histidyl imidazolyl group (residue 51). The N-2 nitrogen of this histidyl group (Fig. 6) extends into aqueous solution (47). This histidyl imidazolyl group is completely isolated from both clefts by intervening amino acid residues and, therefore, is unlikely to be involved in catalysis *via* direct bonding interactions to either coenzyme or substrate. *Brändén et al.* (47) suggest that the $Zn^{2+}-OH_2 \ldots OH(Ser-48) \ldots N(His-51)$ system may function as a "proton charge relay" as shown in Eq. (24).

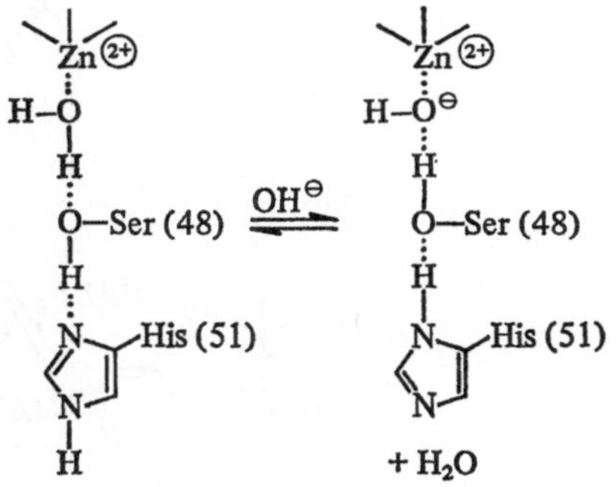

$$
\begin{array}{c}
\searrow\overset{|}{\underset{H-O}{Zn}}^{(2+)} \\
\vdots \\
H \\
\vdots \\
O-Ser\,(48) \\
\vdots \\
H \\
\vdots \\
N\!\!-\!\!His\,(51) \\
\end{array}
\quad\xrightleftharpoons{\;OH^{\ominus}\;}\quad
\begin{array}{c}
\searrow\overset{|}{\underset{H-O^{\ominus}}{Zn}}^{(2+)} \\
\vdots \\
H \\
\vdots \\
O-Ser\,(48) \\
\vdots \\
H \\
\vdots \\
N\!\!-\!\!His\,(51) \\
\\
+\,H_2O
\end{array}
\tag{24}
$$

This suggestion offers a plausible explanation for the afore-mentioned pH-dependencies of NAD^+ and NADH binding to the native enzyme, as shown in Eq. (25), if it is assumed that the ionization of the water molecule affects the site affinity for coenzyme (47).

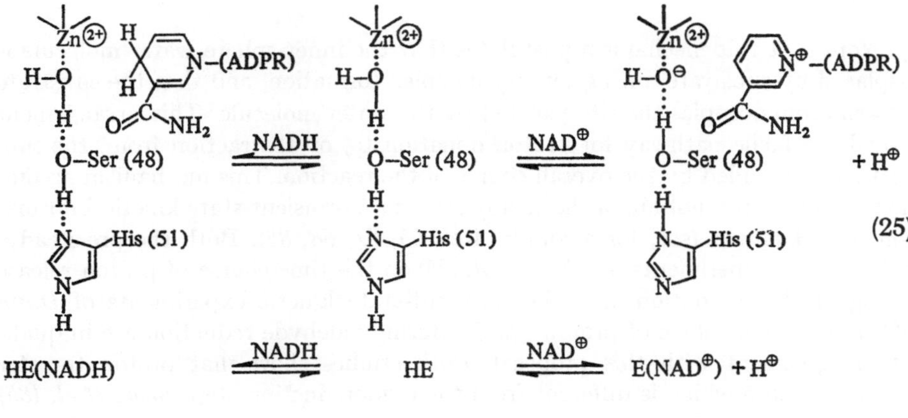

$$
\text{HE(NADH)} \;\xrightleftharpoons{\;NADH\;}\; \text{HE} \;\xrightleftharpoons{\;NAD^{\oplus}\;}\; \text{E(NAD}^{\oplus}) + H^{\oplus}
\tag{25}
$$

Since there appear to be no other ionizable groups in close proximity to the presumed nicotinamide binding site, this explanation appears to be highly likely. *Brändén* further proposes that this "charge relay" system plays an acid-base catalytic role in the chemical transformation. A mechanism consistent with this suggestion is shown in Scheme II.

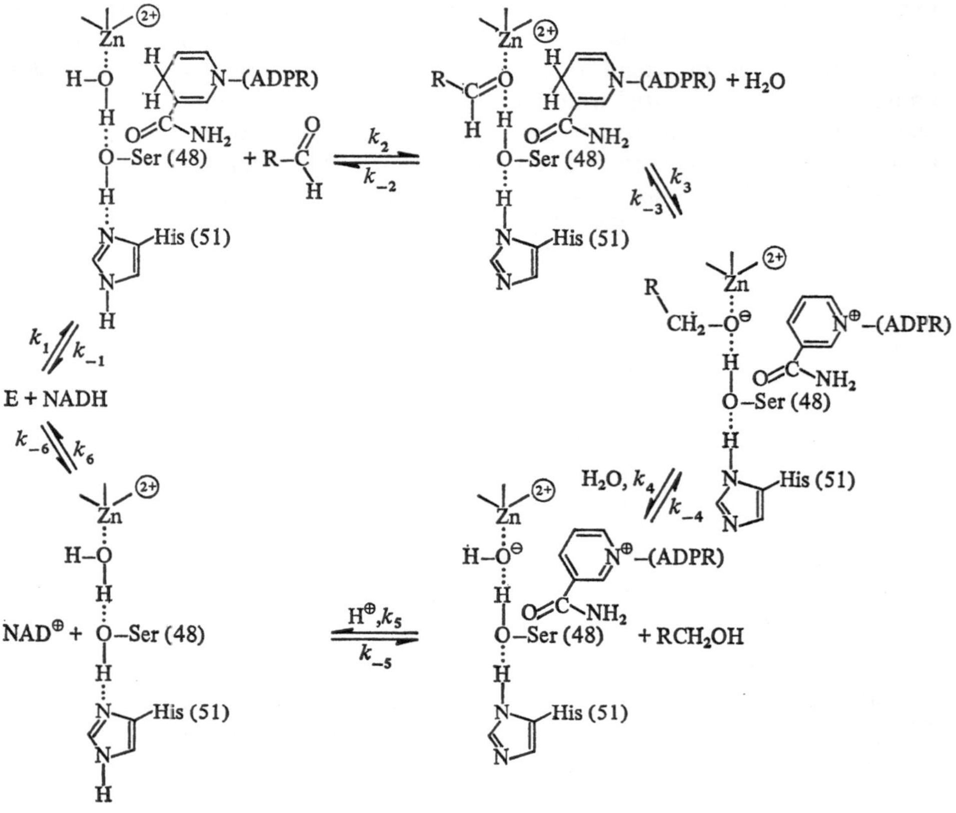

Scheme II

Note that this mechanism postulates that the inner sphere water molecule is displaced by substrate during ternary complex formation, and that the substrate oxygen atom occupies the site vacated by the water molecule. This arrangement provides a facile pathway for proton donation to, or abstraction from, the substrate as demanded by the overall course of the reaction. This mechanism accommodates most, but not all, of the steady-state and transient-state kinetic information on pH effects (see, for example, Ref. 65, 67, 68, 85). Both the pre-steady-state kinetic experiments of *Shore et al.* (85) on the time-course of proton release during alcohol oxidation, and the transient-state kinetic experiments of *Dunn* (68) on the time-course of proton uptake during aldehyde reduction are in qualitative agreement with this proposal. Both studies show that proton transfer occurs in a step which is different from the oxidoreduction step. *Shore et al.* (85)

found that the pre-steady-state burst appearance of NADH observed during the oxidation of ethanol (as discussed above), is preceded by a still more rapid burst release of protons from the enzyme site. The pH-dependence of the amount of hydrogen ion released is shown in Fig. 7. *Dunn (68)* has found that, under the single-turnover conditions of $[NADH]_0 > [E]_0 > [S]_0$, proton uptake during the reduction of 4-(2'-imidazolylazo)-benzaldehyde at pH 8.8 occurs at the (slow) rate of product dissociation from the site, although the time-course of the aldehyde optical-density changes show that a net transformation of substrate to product occurs on a much faster time scale. *Shore et al. (85)* propose that H^+ release from the $E(NAD^+, alc)$ complex (Fig. 7) occurs via ionization of $Zn(II)$-coordinated H_2O, not the ionization of alcohol (*viz.* Scheme II). Furthermore, they propose a general acid-base catalytic role for this H_2O molecule in the oxidoreduction step.

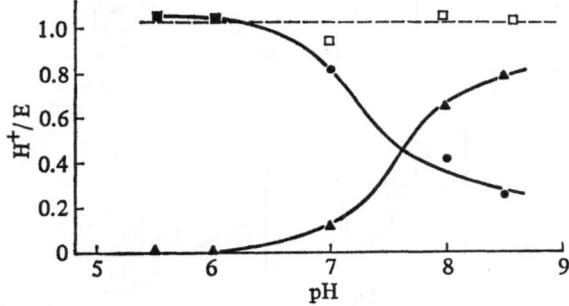

Fig. 7. pH dependence of proton liberation in the formation of alcohol dehydrogenase binary and ternary complexes. Concentrations of enzyme, NAD^+ and trifluoroethanol were 100 μN, 2 mM and 0.5 mM respectively at pH values of 7.0 to 8.5 and 100 μN, 5 mM and 10 mM at pH values below 7. ▲, proton release due to NAD^+ binding; ●, additional proton release due to trifluoroethanol binding; □, additive protons due to ternary complex formation. Taken from Ref. *(85)* with permission

If, as suggested in Scheme II, the inner-sphere coordinated water molecule is displaced by substrate in the ternary complex, then, as discussed in Section II-2, dissociation of coordinated H_2O (assuming either an S_N1 mechanism or an addition-elimination mechanism) places an upper limit on the rate at which substitution by substrate can occur. Therefore, it is of particular interest to note that the bimolecular process of intermediate formation observed for DACA, $4-6 \times 10^7 M^{-1} sec^{-1}$ *(72)*, is comparable in magnitude to the apparent second-order rate constants observed for ligand substitution of inner sphere H_2O from the zinc-aquo complex, $\sim 3 \times 10^7 M^{-1} sec^{-1}$ *(2—5)*. The (perhaps fortuitous) similarlity of these rate constants suggests that the rate of dissociation of inner sphere water from an $NADH-E-Zn(H_2O)^{2+}$ complex limits the rate of enzyme-DACA combination.

The detailed mechanism of the redox step (k_3, k_3, Scheme II) has been the subject of considerable speculation *(65, 91—96)*. *Hamilton (95)* has reviewed several possible mechanisms, and he presents a strong case for biological oxidoreduction mechanisms in which "hydride ion" is transferred as $H^+ + 2e^-$ *via* addition-elimination processes rather than as H^- or as $H \cdot + e^-$. Still, other possibilities have been discussed by *Bruice* and *Benkovic* (96) and *Kosower* (93). These mechanisms can be divided into three general classes: (a) simple hydride transfer mechanisms [Eq. (26)], (b) mechanisms involving adduct formation and interconversion within a cyclic transition state [Eq. (27)], and (c) free radical mechanisms [Eq. (28)].

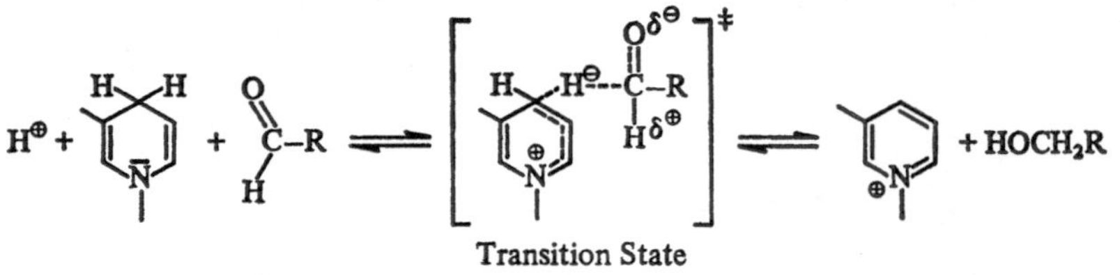

$$\text{(26)}$$

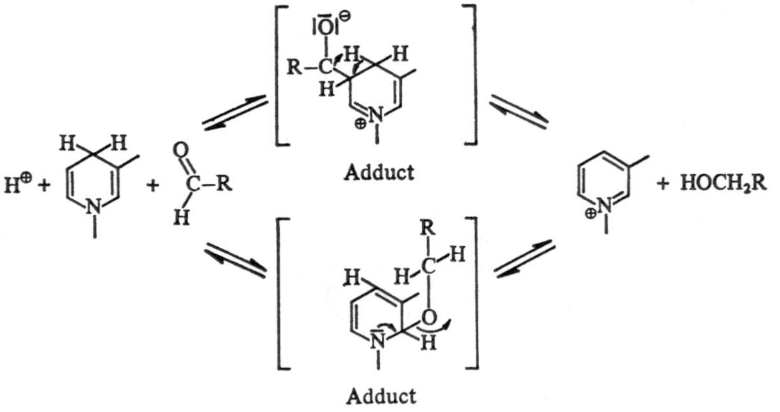

Transition State

$$\text{(27)}$$

Adduct

Adduct

$$\text{(28)}$$

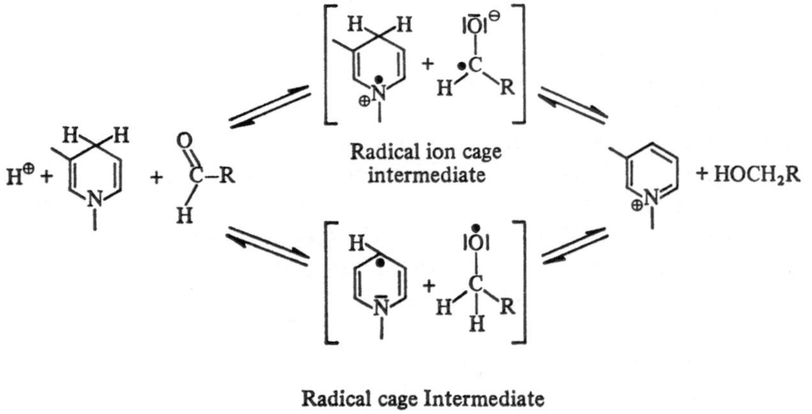

Radical ion cage intermediate

Radical cage Intermediate

M. F. Dunn

It is meaningful to draw a distinction between the mechanisms of Eqs. (26) and (28) only if radical ion pairs such as those shown in Eq. (28) correspond to metastable species along the reaction pathway rather than to a transition-state configuration of reacting species. The deuterium isotope discrimination studies of *Creighton et al. (23)*, see Section III-1, indicate that such species may be important in model systems. The experimental evidence presently available does not rule out any one of these three classes of mechanisms. Indeed, in this author's opinion, an unequivocable statement concerning the mechanism of the redox step will only be possible if a more definitive set of experimental observations is forthcoming.

3. Human Carbonic Anhydrase C

Carbonic anhydrase (E.C.4.2.1.1) is noted for its virtually ubiquitous presence in the tissues of the higher plant and animal species. It is present in particularly large quantities in mammalian erythrocytes where its physiological function includes both the rapid dehydration of bicarbonate in the lungs, and the rapid hydration of carbon dioxide produced as a consequence of metabolism in other tissues (31, 97). The human erythrocyte (97) contains two major forms of carbonic anhydrase, the B form and the C form, both of $\sim$ 30,000 molecular weight (31). The two forms show considerable sequence homology, and each consists of a single polypeptide chain of $\sim$260 amino acids. Both forms have been completely sequenced only recently (98, 99). Both enzymes contain a single, tightly-bound zinc ion which is essential for catalytic activity (100). It has been possible to replace this zinc ion with a variety of divalent metal ions (Table 2), but only the cobalt(II) substituted enzyme retains a catalytic activity which is comparable to, or slightly greater than, that of the native enzyme.

Table 2. Relative activities of divalent metal ion substituted human carbonic anhydrase derivatives[a]

Metal ion[b] derivative	Relative CO_2 hydration activity	Relative rate of p-nitrophenylacetate hydrolysis
Apoenzyme	1.3	1.2
Mn(II) B	1.3	5
Co(II) B	20	120
Ni(II) B	2	4.5
Cu(II) B	0.4	6.9
Zn(II) B	30	37
Zn(II) C	100	100
Cd(II) B	1.4	2.7
Hg(II) B	0.02	1.2

[a] Values calculated from data given in Ref. (30).
[b] B and C refer respectively to the human B and C isozymes.

In addition to the reversible hydration of CO_2 [Eq. (29)]

$$CO_2 + H_2O \;\rightleftharpoons\; HCO_3^{\ominus} + H^{\oplus} \qquad\qquad (29)$$

carbonic anhydrase has been found to catalyze the hydration of various aldehydes (101), to catalyze the hydrolysis of reactive esters such as p-nitrophenylacetate (102—104), and to catalyze the hydrolysis of 2-hydroxy-5-nitro-d-toluenesulfonic acid sultone (105).

In a series of brilliant crystallographic studies, the x-ray crystallography group at the University of Uppsala (40—42, 106) has solved the structure of human erythrocyte carbonic C to a resolution of 2 Å. This group also has given a preliminary report on the structure of the B isozyme, and the full details of this structure should be available in the near future. In addition to the structure of the native C enzyme, the structures of a variety of enzyme inhibitor complexes have been solved (40, 41).

The structure of the carbonic anhydrase active site region (corrected to agree with the recently completed amino acid sequence) is shown in Fig. 8. The notable features of the active site region include, in addition to the distorted tetrahedral ligand field about the zinc ion, an active site cavity which is made up of moderately polar side chain residues. The electron density maps in the region of the active site cavity show peaks of electron density which can not be accounted for as main chain or side chain protein residues. In the various inhibitor complexes, some or all of these densities are displaced depending on the bulk of the inhibitor. *Liljas et al.* (42) have proposed that these densities arise from an ordered, ice-like, water structure involving at least nine water molecules (including the zinc-bound H_2O) which are hydrogen-bonded to each other and to the polar side chain residues (*e.g.*, Gln-66, His-63, Gln-91 and Gln-139) within the cavity as depicted in Fig. 8 A. The hydroxyl group of Thr-197 (see Fig. 8) is located within hydrogen-bonding distance of the inner sphere water molecule at the bottom of the active site cavity. His-63 is located about 6 Å from the zinc ion near the entrance to the cavity (42). Where sequencing information is available for other carbonic anhydrases, both Thr-197 and His-63 are present (31, 98, 99).

The refined coordinates for a number of inhibitors bound to the active site of carbonic anhydrase C are now available, *Bergstén et al.* (41). The inhibitors of direct interest to the following discussion of mechanism fall into two classes: mono-anions (*e.g.*, halides, thiocyanate) and aryl (or aromatic heterocyclic) sulfonamides. Without exception, the donor atoms of both classes of inhibitors replace the zinc-bound water molecule and occupy a position which is within bonding distance of the active site zinc ion. Nevertheless, on the basis of distances found in model systems (see Section II-1) the zinc ion-donor atom distances (Table 3) are somewhat longer than expected, and their lengths decrease on going from pH 8.5 tg pH 7.2 (41). In this context it should be mentioned that the normal bond distance for the inner sphere coordinated water is 2.0 Å (40). In all of the structures investigated thus far, the hydroxyl group of Thr-197 is positioned within hydrogen-bonding distance of the inhibitor. (The hydroxyl oxygen is located 4.5 Å from zinc ion in the native enzyme). This hydroxyl group hydrogen-bonds directly to the donor atom in the anion complexes. The sulfonamide complexes

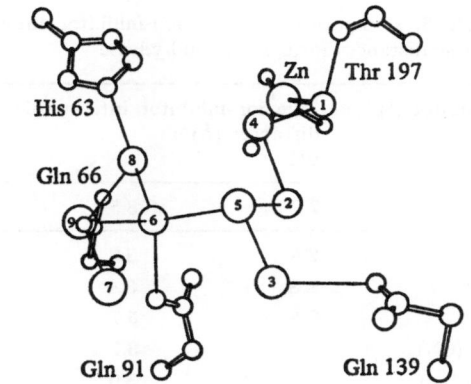

Fig. 8 A

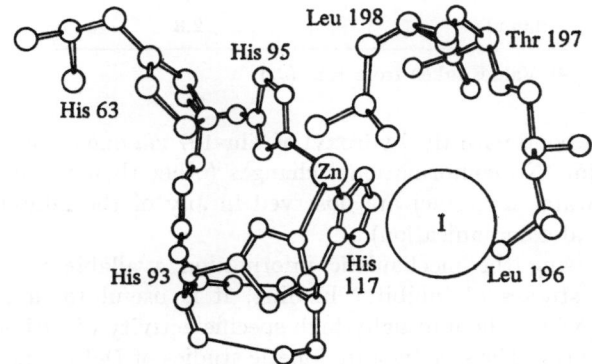

Fig. 8 B

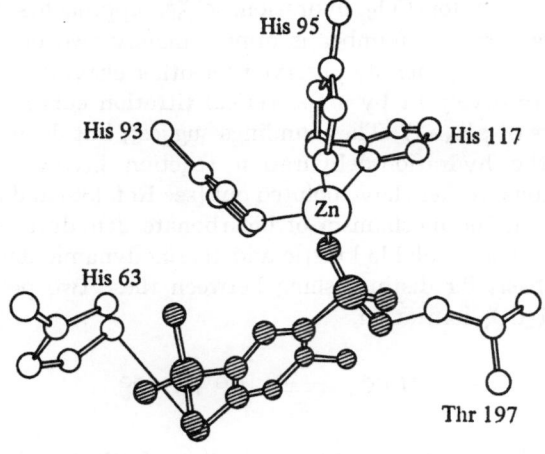

Fig. 8 C

Fig. 8. Three-dimensional representations showing the structural details of the human carbonic anhydrase C active site for: (A) the native enzyme, (B) the enzyme-iodide ion complex, and (C) the enzyme-salamide complex. Taken from Ref. (41) with permission. The residues numbered 1—9 in (A) are believed to be solvent molecules (presumably water molecules). Residue 1 occupies the fourth ligand site in the inner coordination sphere of the active site zinc ion. The atomic positions (solid spheres) of the sulfonamide inhibitor, salamide, are indicated in (C)

Table 3. Comparison of zinc ion-inhibitor inter-
atomic distances for carbonic anhydrase

Inhibitor (I)	Zinc ion-inhibitor interatomic distances (Å)[a] pH	
	7.2	8.5
Cl⁻	2.8	3.3
Br⁻	2.8	3.4
I⁻	3.5	3.7
Au (CN)$_2^-$		3.7
AM Sulf		3.0
Acetazolamide		3.6
Salamide		2.8

[a]) Values taken from Ref. (41).

form a hydrogen bond with the hydroxyl of Thr-197 *via* one of the oxygen atoms attached to sulfur. No conformational changes (other than the displacement of the zinc-bound water molecule) are observed in any of the inhibitor complexes (*Kannan*, personal communication).

Before discussing the mechanistic information available from kinetic and thermodynamic studies of inhibitor binding, it is useful to first consider the severe strictures which the unusually-high specific activity of carbonic anhydrase places on mechanism. The steady-state kinetic studies of *DeVoe* and *Kistiakowsky* (*107*), *Kernohan* (*108*), *Khalifah* (*109*), and *Magid* (*110*) have established that the specific turnover rate for CO_2 hydration, $k_{cat}^{CO_2}$, approaches 5×10^5 sec⁻¹ at neutral pH. This turnover number is approximately two orders of magnitude greater than the values generally observed for other enzymes (*4*). The pH profile for $k_{cat}^{CO_2}$ is quantitatively fit by a theoretical titration curve for a single protic ionization process of $pK_a \sim 7$. These findings suggest, but do not prove, that the mechanism of the hydration-dehydration reaction involves protic acid-base catalysis. As various workers have pointed out [see Ref. (*30*) and (*31*)], the reactive substrate species in the mechanism of bicarbonate dehydration could be either H_2CO_3 or HCO_3^-. The available kinetic and thermodynamic data do not provide an unequivocal basis for distinguishing between these two possibilities because (a) the rate of H_2CO_3 ionization,

$$H_2CO_3 \; \underset{k}{\overset{k_t}{\rightleftharpoons}} \; H^\oplus + HCO_3^\ominus \tag{30}$$

is a relatively rapid process, $k_t \simeq 10^7$ sec⁻¹ (*4*), and (b) over the entire pH range accessible to study, HCO_3^- and CO_2 are the predominate species in solution. Thus, it has not yet been possible to distinguish between the pathways of Eq. (31) and Eq. (32) for the production of bicarbonate.

$$CO_2 + H_2O \; \overset{E}{\rightleftharpoons} \; H^\oplus + HCO_3^\ominus. \tag{31}$$

$$CO_2 + H_2O \; \overset{E}{\rightleftharpoons} \; H_2CO_3 \leftrightharpoons H^\oplus + HCO_3^\ominus \tag{32}$$

These two possibilities have been treated in detail in the recent papers by *Koenig* and *Brown* (*111*), *Khalifah* (*112*), and *Lindskog* and *Coleman* (*113*). The following discussion is based on the arguments presented in these papers. Equation (31) demands that a protonated enzyme intermediate be formed concomitant with the formation of HCO_3^- [Eq. (33)].

$$E + CO_2 \; \rightleftharpoons \; E \cdot CO_2 \; \overset{H_2O}{\rightleftharpoons} \; EH^\oplus \cdot HCO_2^\ominus \; \rightleftharpoons \; EH^\oplus + HCO_3^\ominus \tag{33}$$

This protonated enzyme must undergo ionization to regenerate the active form of the enzyme (E) before the turnover cycle is complete:

$$EH^\oplus \; \overset{k_f}{\underset{k_r}{\rightleftharpoons}} \; E + H^\oplus \tag{34}$$

This transformation must occur at a rate which is greater than, or equal to, the rate of turnover. Since the interconversion of the acidic and basic forms of the enzyme involve an ionization of apparent $pK_a' \sim 7$, the maximum allowable rate of this step can be estimated (after *Khalifah*) from the following considerations. Given the ionization process for the regeneration of active enzyme as denoted in Eq. (34) and assuming that protolysis and hydrolysis make similar contributions at pH 7, k_f must then be $\geq 5 \times 10^5$ sec^{-1} (*i.e.*, the specific turnover number) to accommodate the mechanism of Eq. (31). As a consequence, $k_r \geq 5 \times 10^{12} M^{-1} sec^{-1}$ (*i.e.*, $k_r = k_f/K_E$). This value exceeds the diffusion-limited combination rates observed for small molecule acid-base neutralization reactions in aqueous solution (*4, 114*) by a factor of $\sim 10^2$. As *Khalifah* (*112*) has pointed out, this happenstance can not be circumvented by assuming a large collision radius ("reaction distance") since a rate constant $\geq 5 \times 10^{12} M^{-1} sec^{-1}$ requires a collision radius ≥ 1650 Å. Note that the above estimation of k_r assumes H_2O and/or OH^- are the proton acceptors in Eq. (34). As discussed at length both by *Khalifah* (*112*) and by *Lindskog* and *Coleman* (*113*) (and apparently first suggested by *Dr. W. P. Jencks*), if deprotonation of the enzyme were to occur *via* a general base catalyzed mechanism, then a buffer ion-mediated proton transfer could account for the required rapid proton transfer between enzyme and aqueous milieu.

Since the alternative mechanism, Eq. (32), involves the formation of H_2CO_3 as the enzymic product, the enzyme does not undergo a change in ionization state, and the necessity for a fast proton transfer between enzyme and solution does not arise. *Khalifah* (*109*) has reported that, in agreement with the earlier work of *DeVoe* and *Kistiakowsky* (*107*), the required rate of H_2CO_3 combination with the site, as estimated by the ratio

$k_{cat}^{H_2CO_3}/K_m^{H_2CO_3}$, then would be $\geq 4.7 \times 10^{10} M^{-1} sec^{-1}$

for human erythrocyte carbonic anhydrase C. This value is approximately 10-fold greater than that expected for a diffusion-limited collision between an enzyme site and a neutral substrate (*4, 115*) and, for this reason, it generally has been assumed (*107—110*) that H_2CO_3 can not be the enzymic product in the CO_2 hydration reaction. *Koenig* and *Brown* (*111*) have proposed that the high pH form of the enzyme catalyzes the hydration-dehydration reaction [*via.* Eq. (32)], and that H_2CO_3 is the enzymic reaction product. They suggest (*111*) that the large rate constant estimated for the combination of H_2CO_3 with the site arises from "collisions of H_2CO_3 molecules with the enzyme with subsequent [rapid] surface diffusion to the active site." Since there is little or no precedent in the literature relating to protein-small molecule interactions to support this novel suggestion, it appears more reasonable to conclude that HCO_3^- is the species produced in the enzyme-catalyzed hydration of CO_2. This conclusion should be tempered by the realization that the calculated second-order rate constant for the collision of H_2CO_3 with the catalytic site ($k_{cat}^{H_2CO_3}/K_m^{H_2CO_3}$) is derived from the relationship

$$(k_{cat}^{CO_2}/K_m^{CO_2})/K_h = k_{cat}^{H_2CO_3}/K_m^{H_2CO_3} , \tag{34}$$

where K_h is the equilibrium constant for the reaction $CO_2 + H_2O \rightleftarrows H_2CO_3$. Given the experimental uncertainties inherent in the determination of these parameters (*108, 109*) it could be argued that the above cited bimolecular rate constant is, within experimental uncertainty, no larger than the expected diffusion-limited rate constant. Nevertheless, as will become evident in the following discussion, the binding of mono-anions to the enzyme site is a characteristic property of the enzyme site which appears to be directly related to the catalytic functioning of the active site, and consequently it appears more probable that HCO_3^- is the enzymic reaction product.

Insofar as the mechanism of substrate binding to carbonic anhydrase (*e.g.*, HCO_3^- or H_2CO_3) is likely to parallel the mechanism(s) of anion and/or sulfonamide inhibitor binding, it is of interest to review the recent literature pertaining to the mechanisms of inhibitor binding. The close relationship of the pH-dependence of inhibitor binding to the catalytic and spectral properties of the native and Co(II) enzymes (*32*) is indicative that inhibitor binding is directly relevant to the carbonic anhydrase catalytic mechanism. The pH-dependencies of the affinities of anionic inhibitors for carbonic anhydrase, in virtually all cases, yield an apparent pK_a' for the enzyme ($pK_a' \simeq 7$) which is independent of anion structure (*116—123*). This same pK_a' regulates the activities of both native and Co(II) enzymes in the CO_2 hydration reaction. This pK_a' is also identical to the pK_a' reflected in the pH-dependent visible spectrum of the Co(II) enzyme.

The high resolution X-ray structures of native carbonic anhydrase and carbonic anhydrase-inhibitor complexes (*40—42, 106*) demonstrate that both classes of inhibitors bind to the enzyme site *via* inner sphere coordination to zinc (or cobalt). This finding is in good agreement with spectroscopic studies [see for example (*117, 121—128*)]. Sulfonamide binding to the site, according to *Bergstén et al.* (*41, 106*), involves additional binding interactions in the form of (a) a large number of van der Waals contacts between the sulfonamide aryl moiety and the protein side chain groups which make up a nonpolar region adjacent to the metal ion, (b) hydrogen-bonds to various atoms in the active site including Thr-197 and His-63, and (c) hydrophobic interactions resulting from the displacement of solvent molecules from the active site.

Taylor et al. *(116, 117)* have shown that the pH-dependent affinity of the enzyme for mono-anions arises from a pH-dependent rate of complex formation and an essentially pH-independent rate of complex dissociation. The kinetic investigations of *Taylor et al.* *(116, 117)*, *Kernohan (108)*, *Gerber et al. (118)*, *Thorslund* and *Lindskog (129)*, and *Olander et al. (130)* demonstrate that only the acidic form of the enzyme, EH+, can combine with monoanions to form the inhibitory complex:

$$EH^\oplus + A^\ominus \; \rightleftharpoons \; EH^\oplus - A^\ominus \tag{35}$$

The mechanistic interpretation of the rate and equilibrium data for aryl sulfonamide binding to carbonic anhydrase, gathered by *Taylor et al.* *(116, 117, 119)* and by others *(118, 123, 124, 130)*, is complicated by the ambiguity inherent in the pH-dependence of the apparent association rate constants for complex formation (Fig. 9). *Taylor et al. (119)* found that for several aryl sulfonamides the apparent bimolecular association rate constant (k_a) is highly dependent on pH, while the unimolecular dissociation rate constant (k_d) is virtually independent of pH in the pH range 5.5 to 10.5. The bell-shaped dependence of k_a on pH (Fig. 9) was found to adhere quantitatively to the theoretical curve predicted for a rate process dependent on two protic ionizations. The first ionization was found to correspond numerically to the aforementioned enzyme pK_a' of ~7. The second ionization was found to correspond closely to the ionization constant of the sulfonamide group of the inhibitor. Furthermore, the visible spectrum of the Co(II) enzyme-sulfonamide complex was found to be independent of pH. This fact, as

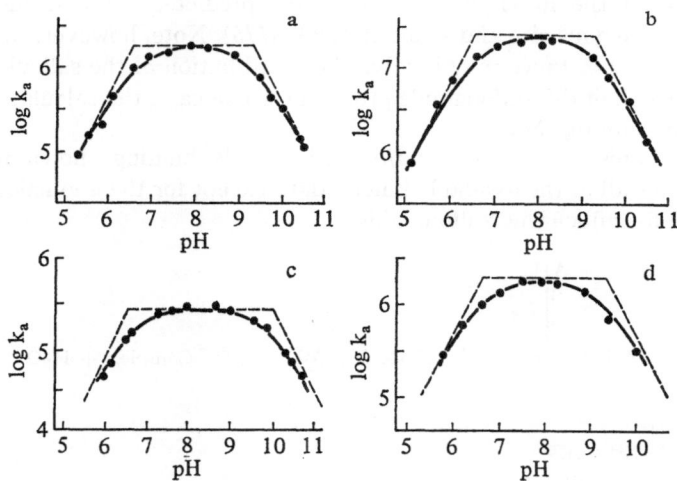

Fig. 9. pH dependence of the association rate constants between various sulfonamides and human carbonic anhydrase C; (a) p-nitrobenzenesulfonamide and Zn isoenzyme C; (b) p[salicyl-5-azo]benzenesulfonamide and Zn isoenzyme C; (c) dansylamide and Zn isoenzyme C; (d) p-nitrobenzenesulfonamide and Co(II) isoenzyme C. The solid lines are calculated assuming pK_E to be 6.60, and pK_S to be 9.30 for p-nitrobenzenesulfonamide and salicylazobenzenesulfonamide and 10.00 for dansylamide. The intersection of the dotted lines with slopes of 1.0 and 0.0 also yield the respective pK values. Taken from Ref. *(119)* with permission

these authors point out, is in accord with the pH-insensitivity of k_d. Thus, either the ionizable group on the native enzyme is displaced during complex formation, or this residue is completely screened from aqueous milieu within the complex and, therefore, is no longer subject to protic ionization. The high resolution 3-dimensional structure of the enzyme-sulfonamide complex (Fig. 8) (41, 106) shows that the water molecule coordinated to zinc ion in the native enzyme is replaced by the sulfonamide group. This finding lends support to the contention that the water molecule is the ionizable group with $pK_a' \sim 7$ at the enzyme site.

Either of the two pathways [Eqs. (36, 37)] shown below will predict the pH-dependence of k_a:

$$E + RSO_2NH_2 \underset{k_{-1}}{\overset{k_1}{\rightleftharpoons}} EH^{\oplus} — NHSO_2R^{\ominus} ; \quad k_a = \frac{k_1}{[1 + K_s/(H^+)] [1 + (H^+)/K_E]} \quad (36)$$

$$K_E \Big\Updownarrow H^{\oplus} \quad H^{\oplus} \Big\Updownarrow K_s$$

$$EH^{\oplus} + RSO_2NH^{\ominus} \underset{k_{-2}}{\overset{k_2}{\rightleftharpoons}} EH^{\oplus} — NHSO_2R^{\ominus} ; \quad k_a = \frac{(k_2 K_s/K_E)}{[1 + K_s/(H^+)] [1 + (H^+)/K_E]} \quad (37)$$

On the basis of the similarities between the kinetic and thermodynamic characteristics of mono-anion binding and sulfonamide binding it seems probable that both mono-anion binding and sulfonamide binding occur by the same pathway, i.e., k_2 [Eq. (37)]. However, Taylor et al. (119) have argued that the values of k_1 and k_2 calculated from the above expressions (Table 4) offer a basis for discarding the possibility that reaction occurs via Eq. (37), since in the case of p-(salicyl-5-azo)benzene-sulfonamide the calculated value for k_2 ($1.12 \times 10^{10} M^{-1} sec^{-1}$) is greater than the maximum rate constant predicted for a diffusion-limited association of enzyme site and small molecule (115). Note, however, that complex formation in this instance may involve the coordination of the salicylcarboxylate moiety rather than the sulfonamide group; in which case, the calculated value for k_2 would be inappropriate.

The mechanism for anion and/or sulfonamide binding shown in Eq. (38) accommodates all of the available kinetic data except for the anomalous behavior of p-(salicyl-5-azo)benzene-sulfonamide.

$$AH$$

$$\Big\Updownarrow K_a$$

$$E—XH + A^{\ominus} \rightleftharpoons E—XH \ldots A^{\ominus} \rightleftharpoons Complex + H_2O \quad (38)$$

$$K_E \Big\Updownarrow$$

$$E—X^{\ominus} + H^{\oplus}$$

It has been proposed that the ionizable group of $pK_a' \sim 7$ is the zinc(II)-coordinated H_2O molecule, as depicted in Eq. (38). If this is not the case, then ligand substitution for water (viz. the X-ray structures, Fig. 8) must alter the apparent pK_a' of the ionizable residue such that protic ionization becomes insignificant in the pH range 5.5 to 10.5. With the possible exception of the Asp—His—Ser hydrogen-bonding system which is postulated to play an integral role in the catalytic

Table 4. Summary of rate constants and equilibrium constants for the reaction of carbonic anhydrase with various sulfonamide inhibitors[a])

Inhibitor	k_a[b]) $(M^{-1}sec^{-1})$	k_1 $(M^{-1}sec^{-1})$	k_2 $(M^{-1}sec^{-1})$	$K_I(M)$	pH	pK_a'	k_{-a} $(sec^{-1}$
4-Hydroxy-3-nitrobenzene sulfonamide[c])	5.48×10^4	1.6×10^5	2.1×10^9	4.17×10^{-6}	6.5	10.9	0.2
Pentafluorobenzene sulfonamide[a])	7.4×10^6	1.2×10^7	2.1×10^{-8}	2.3×10^{-8}	7.6	8.05	0.164
Benzene sulfonamide[c])	1.06×10^5	3.2×10^5	2.5×10^9	1.54×10^{-6}	6.5	10.2	—
p-Nitrobenzene sulfonamide[c])	7.37×10^5	2.2×10^6	7.0×10^8	6.25×10^{-8}	6.5	9.3[d]	0.048
Acetazolamide[c])	4.83×10^6	1.7×10^7	4.4×10^7	5.88×10^{-8}	6.5	7.2[e]	0.068
p-(Salicyl-5-azo)benzene sulfonamide[c])	1.13×10^7	3.4×10^7	1.3×10^7	2.94×10^{-8}	6.5	9.4[d]	0.033
5-Dimethylamino-naphthalene-1-sulfonamide[c])	2.4×10^5	7.2×10^5	1.1×10^9	1.72×10^{-8}	6.5	10.0[d]	0.390
Ethoxzolamide[d])	2.9×10^7	5.5×10^7	1.1×10^9	1.6×10^{-9}	8.0	8.1[d]	0.050
Chlorothiazide[d])	4.5×10^4	5.0×10^4	2.0×10^7	5.4×10^{-7}	8.0	9.4[d]	0.020

[a]) The calculated values of k_1 and k_2 were taken from Ref. (130).
[b]) Experimentally observed second-order rate constants.
[c]) Data taken from Ref. (119).
[d]) Data taken from Ref. (108).
[e]) Data taken from Ref. (129).

mechanism of serine proteases, *Blow et al.* (*131*), *Hunkapiller et al.* (*132*), *Stroud et al.* (*133*), *Robilard* and *Schulman* (*134*), there are no good precedents for the ligand-mediated perturbation of the pK_a' of an enzyme site residue of such a magnitude.

A variety of detailed chemical mechanisms for the carbonic anhydrase-catalyzed hydration of CO_2 have been suggested which appear to be compatible with most, if not all, of the available physico-chemical information [see Refs. (*30—33*)]. The various roles envisaged for zinc ion include: (a) the direct Lewis acid activation of CO_2 *via* inner sphere coordination, (b) enhancement of the nucleophilicity of the attacking water molecule as a consequence of inner sphere coordination and pK_a' perturbation, (c) activation of an outer sphere water molecule for nucleophilic attack *via* base catalysis involving zinc(II)-hydroxide, and (d) a combination of Lweis acid catalysis and nucleophile activation *via* penta-coordinate intermediates (or transition-states). In the following paragraphs the relative merits and the plausibility of select examples of these mechanisms are considered.

A mechanism involving zinc ion as a Lewis acid in the activation of CO_2 is shown in Eq. (39). This scheme accounts for the pH-dependence of carbonic anhydrase activity in terms of the 3-dimensional X-ray structure by invoking the participation of His-63 as an acid-base catalyst in the activation of the attacking water molecule.

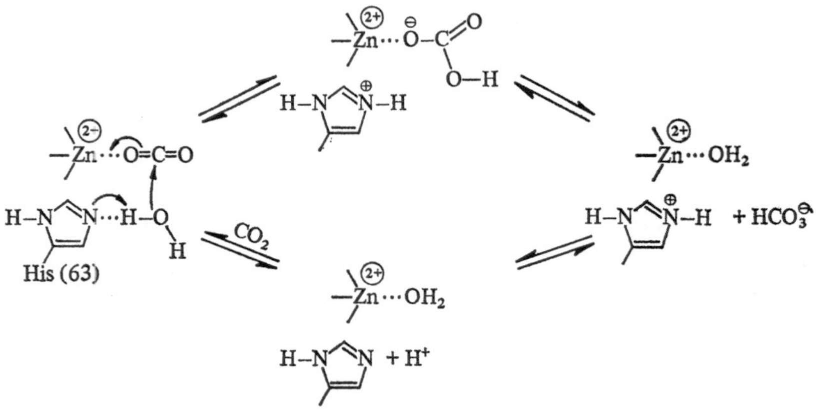

$$\text{(39)}$$

Three arguments can be offered in opposition to this mechanism. They are: (1) The IR studies of *Riepe* and *Wang* (*125*) indicate that CO_2 is not coordinated to zinc ion in the enzyme-CO_2 complex, and CO_2 does not perturb the spectrum of the Co(II) enzyme (*121*). (2) Alkylation of His-63 with bromopyruvate (*135*) only brings about a 70 percent decease in the activity of human carbonic anhydrase C. Therefore, the chemical integrity of His-63 is not crucial to catalytic activity. (3) It is unlikely that anion or sulfonamide binding could sufficiently perturb the pK'_a of His-63 to account for the ionization properties of these complexes, and it is not apparent from the X-ray structure of the Co(II) enzyme (*41, 106*) how the ionization of His-63 can account for the pH-dependence of its visible cobalt(II) spectrum.

These arguments can be criticized on the following bases: (1) The IR studies of *Riepe* and *Wang* and the visible absorption studies of *Lindskog* (*121*) only establish that the most stable enzyme-CO_2 species does not involve inner sphere coordination of CO_2. As pointed out in Section IV-2, such a finding does not rule out the possibility that catalysis involves the obligatory formation of a (high energy) coordination bond between zinc ion and CO_2. Indeed, the direct observation of such a species seems improbable in view of the predictable instability of such a coordination complex. (2) The retention of catalytic activity (albeit greatly reduced) on alkylation of a catalytically essential imidazolyl residue has precedent in the methylation of His-57 in α-chymotrypsin (*136*). Thus, retention of some catalytic activity on alkylation of His-63 does not provide conclusive evidence against the involvement of His-63 as an acid-base catalyst.

Mechanisms in which zinc ion plays a role in the activation of water for nucleophilic attack on CO_2 have dominated the discussion of mechanism in the recent literature. Activation of water is postulated to occur either as a result of the enhanced nucleophilicity of an inner sphere water molecule, Eq. (40), or as a result

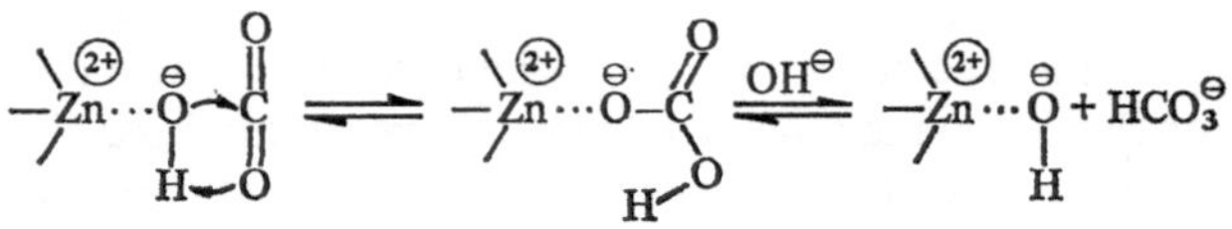

$$\text{(40)}$$

of the participation of a coordinated hydroxide ion as a base catalyst in the activation of an outer sphere water molecule, Eq. (41).

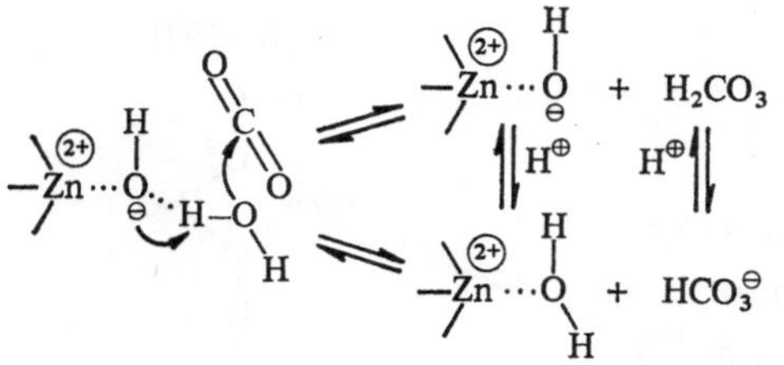

$$\tag{41}$$

It should be noted that hydration mechanisms which proceed *via* the involvement of inner sphere coordinated OH⁻, as pointed out by *Coleman* (*30, 137*) and others (*31, 32, 105, 106, 112, 113, 125, 138*), are attractive in that such mechanisms are compatible with the above discussed anion and sulfonamide studies, the pH-dependent spectral properties of the cobalt(II) enzyme, the pH-dependence of catalytic activity, and the spectroscopic studies of *Riepe* and *Wang* (*125*) and *Lindskog* (*121*).

These mechanisms can be criticized on two points: (1) An abnormally low pK_a' must be assigned to the Zn(II)-coordinated water molecule. (2) The respective step in each mechanism which involves regeneration of the active enzyme species must proceed (in either the forward or reverse direction) at a rate which exceeds the diffusion limit in order to satisfy the kinetic requirements imposed by the high turnover number of the C isozyme.

The abnormally low pK_a' proposed for the carbonic anhydrase Zn(II)-coordinated water molecule ($pK_a' \simeq 7.0$) could arise as a consequence of the unique ligand field about zinc and the microenvironment at the enzyme site. Note that in contrast to alcohol dehydrogenase and carboxypeptidase the carbonic anhydrase zinc ion is bound to the protein *via* three neutral ligands (histidyl imidazolyl moieties). If this pK_a' is compared to the pK_a of the zinc aquo ion ($pK_a \simeq 9.1$, Section II-3, Table 1) then the pK_a' of the enzyme Zn(II)-coordinated H_2O must be perturbed by approximately two pK_a' units. Protein-mediated pK_a' perturbations of this magnitude are not uncommon [see for example (*139—141*)]. Nevertheless, if this ionization is assigned to the coordinated H_2O molecule, then it is difficult to explain why the coordinated H_2O molecules in the Zn(II) and Co(II) enzymes have identical pK_a''s, while the corresponding aquo ions differ by ~0.7 pK_a' units (9.13 ± 0.06 *vs.* 9.85 ± 0.05, see Fig. 9, Section II-3, and Table 1). Arguments pertaining to the second point have been reviewed in preceding paragraphs and will not be further discussed here. Since the involvement of a zinc(II) coordinated hydroxide ion as the active species in CO_2 hydration provides an explanation for most, if not all, of the physico-chemical properties of the enzyme which appear to be related to catalytic activity, this mechanistic postulate is quite attractive.

As pointed out in Section III, it is quite likely that the metal ion has more than one function. Thus, a hydration mechanism involving a penta-coordinate intermediate or transition-state in which both Lewis acid catalysis by direct coordination of CO_2 and nucleophile activation *via* perturbation of the pK'_a of the coordinated water molecule occur can be envisaged [Eq. (42)].

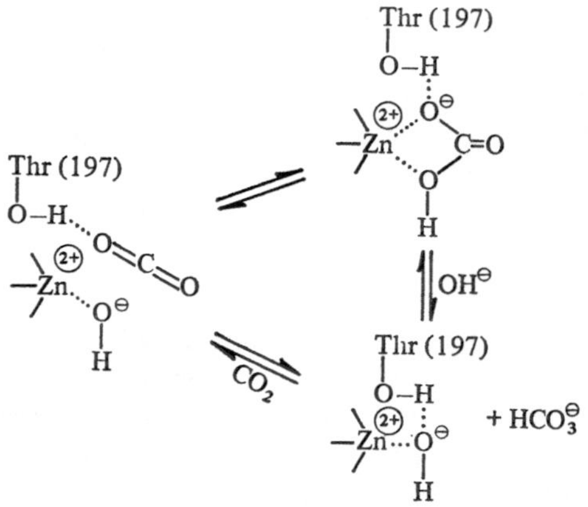

$$(42)$$

The high resolution X-ray structures of the native enzyme and the enzyme-inhibitor complexes (*41, 106*) indicate that Thr-197 plays a role in stabilizing coordination of ligands to the fourth ligand site through hydrogen-bonding between the Thr-197 hydroxyl group and the ligand (Fig. 8). This finding suggests that Thr-197 plays a similar role during the catalytic transformation, as shown in Eq. (42), and a similar role could be suggested for Thr-197 in the above mechanisms [Eqs. (39 to 41)].

In summary, mechanisms which postulate the direct involvement of a Zn(II)-coordinated hydroxide ion in the catalytic mechanism appear to be the most plausible. However, a more definitive statement of mechanism for carbonic anhydrase will not be possible until the role played by the coordinated water molecule is firmly established.

4. Bovine Carboxypeptidase A

Carboxypeptidase A (E.C.3.4.2.1) is one of several proteases secreted by the pancreas as an inactive zymogen, procarboxypeptidase. In contrast to the serine proteases, *e.g.*, chymotrypsin, trypsin and elastase, the zymogen for carboxypeptidase is a zinc metalloprotein consisting of 3 subunits (MW ~87,800). This oligomer is activated to carboxypeptidase A in a complex sequence of events which include the specific proteolysis of approximately 64 amino acids from the N-terminus of one subunit, and dissociation to protomers (*34, 142*). Four major forms of active bovine carboxypeptidase A (α, β, γ and δ) have been isolated. They differ in length by a few amino acids (300 to 307) as a result in differences in the position of bond scission and the degree of subsequent proteolysis during activation. Nevertheless, they all retain similar activities and specificities (*33, 34*).

The general properties, substrate specificity, and the high resolution X-ray structure of carboxypeptidase A have been thoroughly reviewed by *Hartsuck* and *Lipscomb* (*33*) and by *Quiocho* and *Lipscomb* (*34*), and the interested reader is referred to these works and the references contained within for a comprehensive review of the structure and function of carboxypeptidase A. The following discussion presents a highly selected review of the extensive literature that pertains to the catalytic activity and specificity of carboxypeptidase A which, in the author's judgment, is useful to the following discussion of catalytic mechanism.

Carboxypeptidase A is an exopeptidase which specifically hydrolyzes C-terminal aromatic and branched chain aliphatic amino acids from di- and polypeptides. Dipeptides in which the N-terminal amino group is free are hydrolyzed only slowly, whereas the corresponding N-acylated dipeptides are rapidly hydrolyzed. The presence of an N-methyl group on the α-amino nitrogen of either the first, or the second amino acid residue of a polypeptide greatly suppresses the rate of hydrolysis (33, 34). *Hanson* and *Smith* (143) and *Abramowitz et al.* (144) have shown that the identity of the side chain residue in polypeptides of up to five amino acid residues in length influences the magnitudes of K_m and k_{cat}.

Carboxypeptidase also catalyzes the hydrolysis of esters (*i.e.*, O-acyl derivatives of L-β-phenyllactic acid and L-mandelic acid), (145—148). Apparent differences in esterase *vs.* protease activities of chemically modified enzymes have led to the suggestion that there are gross differences in the catalytic mechanisms for these two reactions (149).

As often noted (33, 34, 145—151) the kinetic behavior of carboxypeptidase A is fraught with anomalies which include the observation of substrate activation and/or inhibition, product activation and/or inhibition, and the observation of pH transitions which disappear at high ionic strength. These workers have suggested that most, if not all, of the kinetic anomalies can be rationalized by envoking a multiple loci-substrate binding model (145, 149—151) and/or nonproductive binding modes (33, 146—148). The early work of *Hanson* and *Smith* (143) and *Abramowitz et al.* (144) suggest the presence of multiple subsites for polypeptide binding. More recently, *Davies et al.* (152) and *Auld* and *Vallee* (151) have presented kinetic studies which strongly support this rationalization. They found that α-N-carbobenzoyloxyglycine (CbzGly) activates the hydrolysis of certain dipeptides; e.g., 5×10^{-2}M CbzGly activates the rate of hydrolysis of CbzGly-L-Phe by ~4-fold. In contrast, the rate of hydrolysis of the tripeptide CbzGlyGly-L-Phe is competitively inhibited by CbzGly under similar conditions (151). Furthermore, it was found from structure-reactivity correlations that the steady-state rate of tripeptide hydrolysis (k_{cat}) is relatively insensitive to the nature of the C-terminal residue, but K_m is not. Thus, k_{cat} for the tripeptides CbzGlyGly-L-Phe, CbzGlyGly-L-Leu, and CbzGlyGly-L-Val was found to remain essentially constant, but K_m for these substrates varies by a factor of ~100 (151). However, when the N-terminal residue was varied, holding the C-terminal residue cosntant, k_{cat} was found to vary more than K_m.

Such effects are not unique to carboxypeptidase A. The rate of substrate polysaccharide hydrolysis by lysozyme is remarkably dependent on the polysaccharide chain length (153, 154). Both steady-state kinetic studies and X-ray crystallographic studies on enzyme-inhibitor complexes for chymotrypsin (156) trypsin (156), elastase (157), and subtilisin (158) are indicative of the existence of multiple-loci substrate binding sites. Furthermore, the dependence of k_{cat} on substrate chain length for all these enzymes strongly implies that the filling of these subsites enhances catalytic activity. The existence of such an effect has been shown more directly for trypsin. *Ingram* and *Wood* (*159*) have found that low (mM) concentrations of N-methyl guanidinium ion enhance the rate of turnover of the nonspecific substrate acetylglycine ethyl ester by nearly an order of magnitude. This observation strongly implies that the weak bonding interactions between the specificity subsite and substrate are important for catalysis. Comparison of the X-ray structure of trypsin and the trypsin-pancreatic trypsin inhibitor complex provides direct evidence for the occurrence of a conformational charge when a positively-charged side chain residue fills the specificity subsite (*156*).

While the available data do not allow a distinction to be made between explanations based on the nonproductive binding hypothesis (*160, 161*), the induced fit hypothesis (*162*), or the strain-distortion hypothesis (*163*), the serine protease X-ray structural studies suggest that the kinetic enhancement attributed to the filling of subsites by polypeptide substrates involves only small differences in protein conformation.

The recent investigation of the pH-dependent relaxation spectrum of carboxypeptidase A *via* temperature perturbation of equilibrium, *French et al.* (*164*), provides independent evidence which is consistent with the above interpretation of pH effects on the catalytic activity of carboxypeptidase A. Through the use of pH-indicator dyes these workers found that, on perturbation of equilibrium in the neutral pH range, the native enzyme gives two well separated relaxations ($1/\tau_1 \simeq 4 \times 10^2$ sec^{-1}, $1/\tau_2 \simeq 4.6 \times 10^3$ sec^{-1} at pH 6.75). Since both of these relaxations are independent of the protein and the indicator concentrations, it was concluded that they originate from isomerizations of the monomeric enzyme. The pH-dependence of τ_1 and τ_2 appear to be coupled respectively to the ionization of an enzyme residue, $pK_a' \simeq 7.2$. Although the relationship of these relaxations to the catalytic mechanism has not yet been established, it is probable that the process which is responsible for τ_1 is also the process which regulates the activity of the enzyme at low pH (see the foregoing discussion), since τ_1 is abolished when inhibitory amino acids are bound to the enzyme.

The tripeptides investigated by *Auld* and *Vallee* (*151*) exhibit pH vs. k_{cat}/K_m profiles (Fig. 10a) which indicate that the ionizations of two enzyme residues, apparent pK_a''s of ~6.1 and ~9.0, control catalytic activity. The plots shown in Figs. 10b and 10c demonstrate that the low pH ionization affects only k_{cat}, while the high pH ionization affects only K_m. These authors point out that the simplest scheme which is consistent with these data involves the existence of three enzyme protonation states (*e.g.*, EH$_2$, EH and E) which are important to the activity of carboxypeptidase A. The first ionization, EH$_2 \rightleftarrows$ EH + H$^+$, generates the catalytically active from of the enzyme without altering the affinity of the enzyme for substrate.[2]

The absence of significant activation or inhibition phenomena by substrates or products when synthetic tripeptides are employed as substrates (*33a*) has led *Auld* and *Vallee* (*151*) to suggest that, since the kinetic behavior of these

---

[2] The observation that only the magnitude of k_{cat} depends on this ionization strongly indicates that $K_m = K_s$.

M. F. Dunn

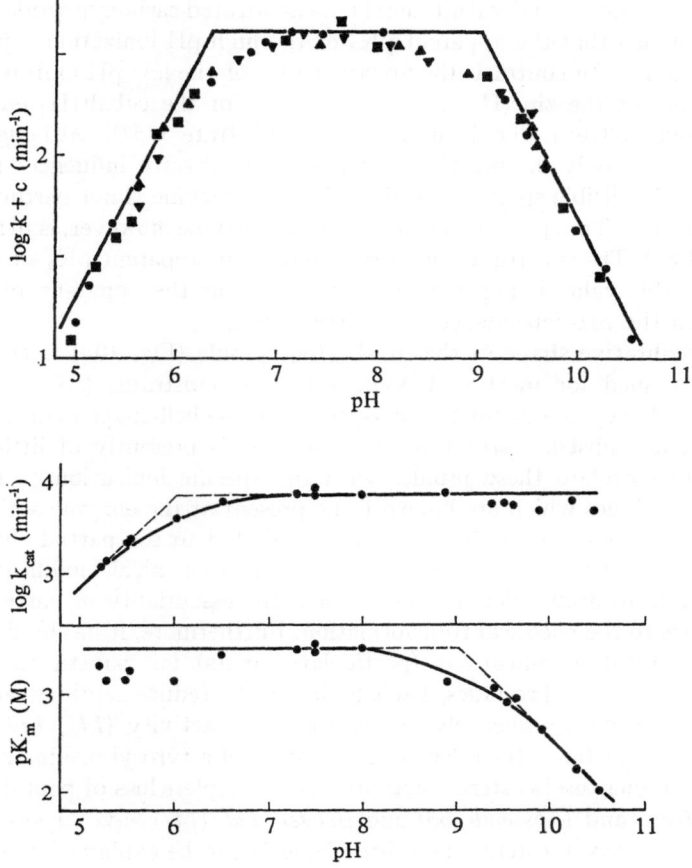

Fig. 10. (A, upper plot) The pH dependence of the hydrolysis of 5×10^{-4} M chloroacetyl-L-phenylalanine ($\blacktriangledown$, $c = 0.25$), 2×10^{-4} M BzGly-L-Phe ($\blacktriangle$, $c = -0.5$), BzGlyGly-L-Leu ($\bullet$, $c = 1.0$), and CbzGlyGlyGly-L-Phe ($\blacksquare$, $c = 0$). Conditions of assay were 1.0 M NaCl, 0.05 M Mes, Hepes, Tris, or carbonate buffer, 25°. The solid lines have slopes of 1, 0, and —1, their intercepts yielding pK_a' values of ~6.1 and ~9.0. (B, C, lower plots) The pH dependence of log k_{cat} and pK_m for CbzGlyGly-L-Phe hydrolysis. Conditions of the assays were the same as in (A)

compounds is more easily interpreted, these substrates are more suitable for the investigation of mechanism than are dipeptides or esters. This argument very likely is valid insofar as the interpretation of activation and inhibition in relation to the catalytic mechanism is concerned. Nevertheless, since the dependencies of k_{cat} and K_m on pH vary as a function of substrate structure, it is quite possible that this variation arises as a consequence of changes in the nature of the rate-limiting step as a function of substrate structure. Therefore, care must be exercised in the extension of this argument to the analysis of pH effects. Similar pH effects have been well characterized as resulting from a change in rate-limiting-step in the instance of amide *vs.* ester hydrolysis by α-chymotrypsin (*165*).

The investigation of divalent metal ion-substituted carboxypeptidase A derivatives has shown that the apparent pK'_a of the high pH ionization is independent of the metal ion. In contrast, the apparent pK'_a of the low pH ionization changes from ~6.33 for the zinc(II) enzyme to ~5.57 for the cobalt(II) enzyme when α-N-benzoyl-GlyGly-L-Phe is used as the substrate (151). Although this pH dependence strongly implies that the metal ion directly influences the low pH ionization, the visible spectrum of the cobalt(II) enzyme is not perturbed by this ionization (166). The spectrum of the cobalt(II) enzyme, however, is pH-dependent above pH 8.0. The spectral changes titrate with an apparent pK'_a of ~8.8 (166). Note that this value is approximately the same as the apparent pK'_a which is reflected in the pH-dependence of K_m (Fig. 10c).

The qualitative shape of the pH-k_{cat}/K_m profile (Fig. 10a) is typical of the profiles obtained for most carboxypeptidase A substrates (33). However, (as mentioned above), the quantitative aspects of these bell-shaped curves are highly dependent on substrate structure. Therefore, it is presently of little utility to attempt to correlate these profiles with the specific ionization of the various functional residues which are known to be present at the enzyme active site.

Chemical modification studies have contributed to the partial elucidation of two aspects of the catalytic mechanism: Through chemical modifications it has been possible to draw inferences concerning the essentiality of particular active site residues to the chemical transformation. Furthermore, it has been shown that certain chemical modifications (particularly metal ion substitutions and the modification of tyrosyl residues, Table 5) drastically reduce peptidase activity and, at the same time, significantly increase esterase activity (145, 149, 167—169). Indeed, in one instance the selective acetylation of a tyrosyl residue brings about both a 7-fold increase in esterase activity and a complete loss of peptidase activity (40). *Hartsuck* and *Lipscomb* (33) and *Bender et al.* (147) have suggested that the observed increases in esterase activity (Table 5) can be explained as resulting (in part) from the alleviation of substrate inhibition. Note however that this argument does not explain how these chemical modifications can occur to give an enzyme species which retains esterase activity but loses protease activity.

Hartsuck and *Lipscomb* (33) report that the high resolution X-ray structural studies indicate that peptide substrates do not interact with the metal ion of enzyme modified by the acetylation of tyrosyl residues. This finding is in agreement with the observations of *Coleman et al.*(170) which show that peptide substrates do not inhibit metal ion exchange in the acetylated enzyme, although metal ion exchange in the native enzyme is strongly inhibited.

More recently, *Kang* and *Storm* (171) have shown that conversion of cobalt(II) carboxypeptidase A to the corresponding cobalt (III) enzyme by hydrogen peroxide oxidation is accompanied by the complete loss of peptidase activity when assayed with CbzGly-L-Phe. Nevertheless, the esterase activity, as measured by the rate of hydrolysis of O-(N-benzoyl-Gly)-D,L-phenyllactic acid, remains unaffected by this transformation. Cobalt(III) complexes are characterized by extremely slow exchange rates due to the transformation from a d^7 (CoII) to a d^6 (CoIII) electronic configuration, (see Section III-2). Therefore, *Kang* and *Storm* conclude that either peptides and esters are hydrolyzed by different mechanisms, or that hydrolysis occurs *via* a common mechanism which does not involve an

Table 5. Comparison of peptidase and esterase activities of various carboxy-peptidase A derivatives[a]

Enzyme derivative	Relative Peptidase activity(%)[d]	Relative esterase activity(%)[e]
(A) Covalent modification[b] (Tyr. Residues Modified)		
Acetylation (Tyr 248, 198)	0	700
Mild Diazo Coupling (Tyr 198)	100	200
Strong Diazo Coupling (Tyr 248, 198)	0	200
Nitration (Tyr 248, 198)	10	170
(B) Metal Ion Substitution[c]		
Zn(II)	100	100
Co(II)	160	95
Ni(II)	106	87
Mn(II)	8	35
Cu(II)	0	0
Hg(II)	0	116
Cd(II)	0	150
Pb(II)	0	52

[a] This table is a composite of tables found in Ref. (*33*).
[b] Values have been calculated from the original data contained in Refs. (*145, 167, 168*).
[c] Values have been calculated from the original values contained in Ref. (*172*).
[d] The peptide substrate used was N-carbobenzoyloxyglycyl-L-phenyl-alanine.
[e] The ester substrate used was O-hippuryl-L-β-phenyllactate.

inner sphere coordination bond to the metal ion. In this context, it is important to note that substitution of either Cd(II) or Hg(II) for zinc ion yields modified enzymes with virutally no peptidase activity when assayed with the peptide substrate N-CbzGly-L-Phe, Table 5. However, both derivatives retain high esterase activities when assayed with O-hippuryl-L-β-phenyllacetate (*172*). These experiments should be interpreted with caution since the recent work of *Zisapel* and *Sokolovsky* (*173*) on the bovine carboxypeptidase B enzyme (a closely related exopeptidase) suggests that the presence or absence of peptidase activity for the corresponding Co(II) and Hg(II) derivatives of this enzyme is critically dependent on the length of the polypeptide substrate. For this reason, it would be informative to test the peptidase activities of the Cd(II) and Hg(II) enzymes with a variety of polypeptide substrates, particularly tri- and tetrapeptides.

The results of the high resolution X-ray diffraction studies on crystalline carboxypeptidase A have received considerable attention. This series of elegant 3-dimensional structural determinations by *Lipscomb* and co-workers (*36—39*) have provided a detailed (2 Å resolution) atomic structure for the native enzyme which has been noted for its unusual clarity.

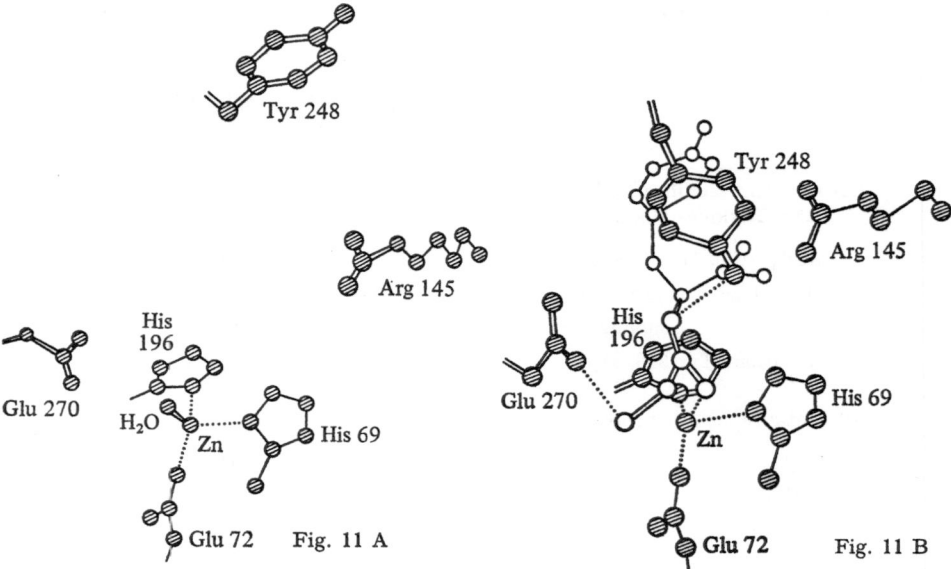

Fig. 11. Three-dimensional representation showing the structural details of the Bovine carboxypeptidase Aγ active site for (A) the native enzyme, and (B) the enzyme-Gly-L-Tyr complex. Note: the large change in the location of Tyr-248 which occurs on complex formation, the coordination bond between the carbonyl group of the substrate amido group and the zinc ion, the salt bridge formed between Arg-145 and the C-terminal carboxylate ion of the substrate, and the salt bridge formed between Glu-270 and the N-terminal α-ammonium ion of the substrate. [Redrawn from Ref. (39) with permission]

The active site zinc ion (see Fig. 11) is located in a shallow cleft which runs across the surface of the protein to the opening of a pocket in the enzyme surface. Note that the three zinc ligands contributed by the protein are positioned away from the cleft, while the inner sphere water molecule projects out into the cleft. The cleft is lined with several highly polar residues in the vicinity of the metal ion which are believed to participate in substrate binding and/or catalysis. The pocket adjacent to the metal ion is lined with nonpolar (hydrophobic) residues. Both the cleft and the pocket are filled with water molecules which appear to be arranged in an ice-like array.

The 2 Å resolution difference electron density map which compares the native enzyme structure with the enzyme-Gly-L-Tyr complex (37, 39) shows extensive differences in conformation involving the positions of three amino acid side chain residues, Arg-145, Glu-270 and Tyr-248 (Fig. 11). The difference map also shows evidence for smaller conformational differences in the region of the active site which involve limited repositioning of the polypeptide backbone and disruption of four probable hydrogen bonding interactions which form a network between Arg-145 and Tyr-248 via residues 155, 154, and 249 (33, 37, 39). These conformational changes are postulated to be necessary in order to optimize the weak bonding interactions between the enzyme surface and the substrate. As *Hartsuck* and *Lipscomb* point out (33), complex formation changes the cleft from an aqueous environment to a cleft within which specific, highly directional weak bonding interactions (hydrogen bonds, Coulombic force fields, London dispersion force fields) occur. The hydrophobic pocket is similarly transposed from an aqueous environment to a nonpolar environment by the insertion of the aryl moiety of the substrate.

The prominent weak-bonding interactions which have been deduced from the 2 Å structure (Fig. 11) include: (a) A Coulombic charge-charge attractive interaction between the C-terminal carboxylate group of the substrate and the guanidinium ion moiety of Arg-145. Arginine-145 undergoes a 2 Å movement to accommodate this interaction. This interaction is interpreted as providing the trigger for the conformational changes that result in the disruption of the hydrogen bonding system which maintains the position of the phenolic residue of Tyr-248 in the native enzyme. As a consequence, Tyr-248 takes up a new position some 12 Å away. This movement allows the formation of a (probable) hydrogen bond between the phenolic hydroxyl group and the amido nitrogen of the substrate. (b) A Coulombic charge-dipole interaction between the carbonyl oxygen of the peptide linkage of the substrate and the active site zinc ion. This interaction occurs *via* the displacement of the zinc-bound water molecule. In this context it should be noted that the NMR studies of *Navon et al.* (*173, 174*), *Koenig et al.* (*175*) and *Quiocho et al.* (*39*) show that competitive inhibitors of peptide hydrolysis displace one or more water molecules from the first coordination sphere of the Mn(II) enzyme on binding. (c) A Coulombic interaction mediated by a water molecule between the carboxylate ion of Glu-270 and the N-terminal α-ammonium ion of the substrate. This interaction results in the displacement of the carboxyl group of Glu-270 by about 2 Å from its position in the native enzyme. (d) A hydrogen bonding interaction between the phenolic OH group of Tyr-248 and the amido nitrogen of the substrate, as discussed in (b). The motions of Arg-145, Tyr-248, and Glu-270 appear to occur primarily as a result of side chain motions involving rotation about C-C single bonds.

As pointed out by *Lipscomb* and co-workers (*33, 34*), Gly-L-Tyr almost certainly is bound in an unreactive mode. Nevertheless, information drawn almost exclusively from the structure of this complex has been used by these authors to formulate a hypothesis for the carboxypeptidase catalytic mechanism. The derivation of their hypothesis, in addition to the incorporation of certain of the above structural features, is based on the following considerations: (1) Conversion of the active site cleft from an aqueous environment to a hydrophobic environment on substrate binding provides a driving force for the reaction, (2) The conformational changes which attend the binding of Gly-L-Tyr (and other substrate analogs) brings a group thought to be required for catalysis (Tyr-248) into hydrogen-bonding contact with the bound substrate molecule. (3) The Coulombic interaction between Arg-145 and the C-terminal carboxyl group conveys specificity to substrate binding. (4) The inner sphere coordination of the substrate carbonyl to the active site zinc ion is brought about by the highly specific interactions discussed above. Since the carbonyl oxygen of the substrate amido function is a comparatively poor donor atom, it is unlikely that this bonding interaction is artifactual in nature. (5) Only the carbonyl group of Glu-270, the phenolic hydroxyl group of Tyr-248 and zinc ion are close enough to the amido group of

Gly-L-Tyr to participate directly in catalysis. (6) Consequently, the binding of Gly-L-Tyr is postulated to closely resemble the catalytically active enzyme-substrate complex.

On the basis of these considerations, *Lipscomb et al.* (*33, 34, 36—39*) have proposed the detailed mechanism shown in Scheme III.

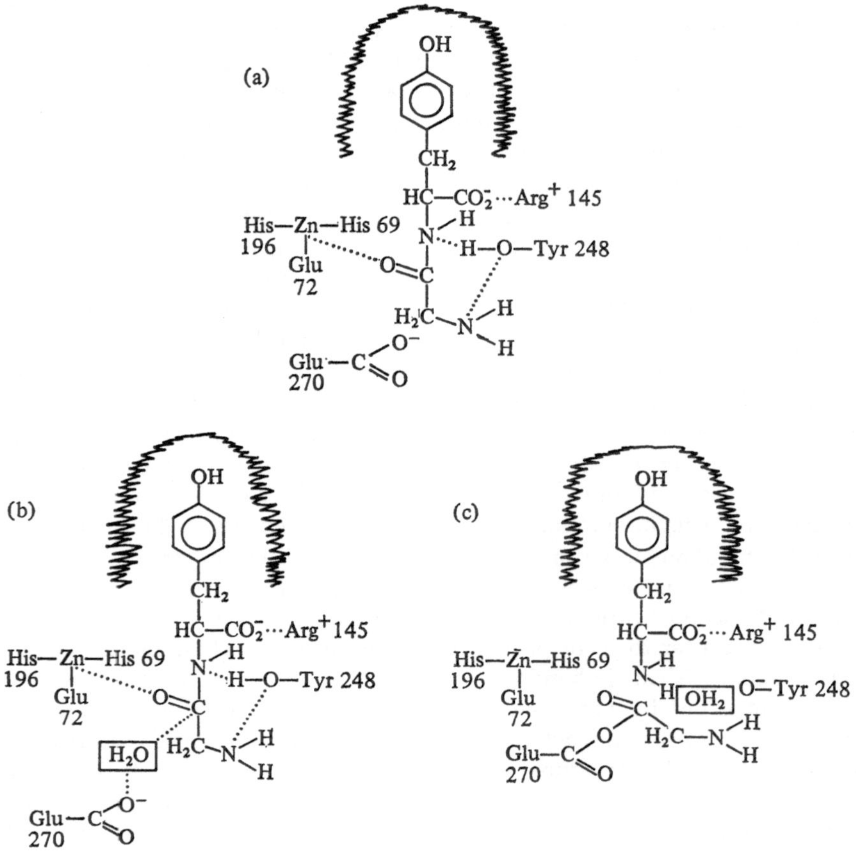

Scheme III

Schematic representation of the possible stages in the hydrolysis of Gly-L-Tyr by carboxypeptidase A. It is probable that the carbonyl carbon of the substrate becomes tetrahedrally bonded as the reaction proceeds, but it is uncertain at what stage of the reaction the proton is added to the NH group of the susceptible peptide bond. (a) Productive binding mode. (b) General base attack by water upon the carbonyl carbon of the substrate. (c) Nucleophilic attack by Glu-270 upon the carbonyl carbon of the substrate. Taken from Ref. (*33*) with permission

The essential features of this mechanism involve the participation of zinc ion as a Lewis acid catalyst which plays an integral role in the process of bond scission by polarizing the carbonyl group of the substrate, thus increasing its susceptibility to nucleophilic attack. The phenolic hydroxyl group of Tyr-248 plays a

M. F. Dunn

(general) acid catalytic role in the protonation of the amido nitrogen to assist bond cleavage by decreasing the bond order of the nitrogen-carbonyl carbon bond, thus aiding the departure of the amino group. Glutamic acid-270 is proposed to function either as a (general) base catalyst in the activation of a water molecule for nucleophilic attack at the carbonyl carbon, or as a nucleophile which attacks the carbonyl carbon. In the latter instance, nucleophilic attack by Glu-270 leads to the formation of an anhydride linkage between enzyme and substrate. Then in a subsequent step, this intermediate undergoes hydrolysis to regenerate enzyme. Zinc ion could participate in the hydrolysis of the anhydride intermediate either *via* Lewis acid catalysis, or *via* the activation of a water molecule for nucleophilic attack.

The above discussion of mechanism explains the low reactivity of Gly-L-Tyr and other dipeptides with free N-terminal α-amino groups as resulting from Coulombic interaction between the α-ammonium ion of the dipeptide and the carboxylate anion of Glu-270. This interaction is seen to effectively prevent the involvement of Glu-270 in catalysis. N-acylation of the α-amino group, however, should be expected to abolish this interaction and restore reactivity since Glu-270 would no longer be constrained by this salt bridge.

The X-ray structural studies offer strong evidence in support of a Lewis acid catalytic role for the active site zinc ion in peptide hydrolysis. Since the carbonyl group is no doubt a much weaker dipole than H_2O, the initial (ground-state) interaction between the zinc ion and the substrate carbonyl oxygen must be stabilized by the summation of the weak bonding forces between enzyme and substrate. The result is to displace the transition-state of the enzyme-catalyzed reaction (relative to its hypothetical nonenzymatic counterpart) along the reaction coordinate toward the enzyme-substrate complex.

While the X-ray crystallographic studies provide a considerable body of evidence in support of the Lewis acid role proposed by *Lipscomb*, it is still possible to argue that the role of zinc ion is to facilitate the nucleophilicity of an attacking water molecule by inner sphere coordination. However, the geometrical constraints imposed by the active site structure make this suggestion seem unlikely, since it then must be postulated that the entire set of weak bonding interactions seen in the Gly-L-Tyr complex are unrelated to catalysis.

Johansen and *Vallee* (176) and *Quiocho et al.* (177) have carried out experiments both in solution and in the crystal phase with a chemically modified, Tyr-248, enzyme derivative to examine the relationship between the conformational state of the residue and the catalytic mechanism. Derivatization of Tyr-248 with diazotized arsanilic acid to form arsanilazo-Tyr-248 carboxypeptidase A

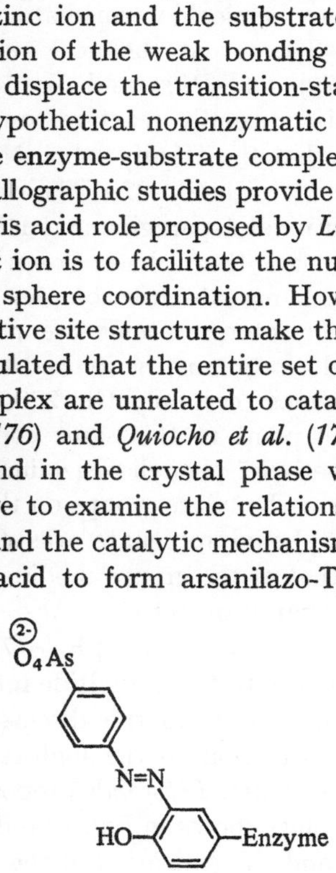

Arsanilazo-Tyr-248 Carboxypeptidase A

112

yields a chromopheric enzyme with spectral properties which are dependent on pH. In basic solution, the azophenol moiety probably complexes with the active site zinc ion to form a chromophoric complex, λ_{max} 510 nm (*176, 177*). The stability of the complex, as measured by its characteristic spectrum, is lost when: (1) the pH is lowered to ~6, (2) chelators such as o-phenanthroline are added, and (3) when substrates or inhibitors are added. Although the spectral changes of this derivative are complicated by the presence of more than two protonation states (*176*), pH-titration studies indicate that the spectrum which arises from the coordination of zinc ion to the chromophore depends on an apparent pK_a' of ~7.7 (*176, 177*). In the crystalline phase, *Quiocho et al.* (*177*) have shown that, at pH 8.2, the spectrum of the chromophore is dependent on the particular crystal form. They found that only the crystal form used in the X-ray structural studies exhibits the 510 nm absorption maximum which characterizes coordination to the active site zinc ion. Although the studies of *Johanson* and *Vallee* (*176*) and *Quiocho et al.* (*177*) differ in some respects with regard to experimental details, experimental results, and interpretation, it is clear from these studies that the arsanilazo-Tyr-248 derivative can undergo a conformational change both in solution and in the crystalline state which is similar to the 12 Å movement of Tyr-248 deduced from the 2 Å resolution structures of the enzyme and the enzyme-Gly-L-Tyr complex (Fig. 11). The relationship of this conformational transition to the catalytic mechanism is open to speculation.

The importance of covalent chemical intermediates in the catalytic mechanism is also open to speculation. The recent chemical modification studies of *Hass* and *Neurath* (*178*), *Pétra* (*179*), and *Pétra* and *Neurath* (*189*) indicate that the carboxyl group of Glu-270 is unusually reactive toward electrophilic reagents. *Hass* and *Neurath* (*178*) find that the affinity label, N-bromo-acetyl-N-methyl-L-phenylalanine reacts specifically with Glu-270. The only other residues modified by this reagent are the N-terminal asparagine and His-13 (in less than stoichiometric amounts). pH studies show that a basic group with an apparent pK_a' ~7 within the reversible formed enzyme-inhibitor complex is involved in the inactivation reaction. Although other interpretations are possible, *Hass* and *Neurath* argue that since the carboxyl group must be ionized to react with the alkylating reagent the basic group is almost certainly Glu-270. The studies of *Petra* (*179* and *Petra* and *Neurath* (*180*) point to this same conclusion. They report that Woodward's reagent K (N-ethyl-5-phenylisoxazolium-3-sulfonate) reacts with Glu-270 as well as with other carboxylate groups on the enzyme. The presence of the competitive inhibitor 3-phenylpropionate was found to protect Glu-270. Furthermore, it was shown that of the groups modified, only the modification of Glu-270 results in the loss of catalytic activity. Also, reaction with Glu-270 was also shown to be dependent on an ionization of pK_a' ~7. Thus, the unusual nucleophilcity of Glu-270 is consistent with the anhydride intermediate mechanism.

Lipscomb (*181*) has presented a speculative discussion of the effects of pH on carboxypeptidase activity in relation to the high-resolution three-dimensional structures. In this interesting paper, *Lipscomb* proposes that the apparent pK_a values observed in the pH profiles shown in Fig. 10 reflect respectively the ionization of Glu-270 (pK_a' ~6.7) and the ionization of the zinc ion-coordinated water molecule (pK_a' ~8.9).

M. F. Dunn

To date, there is no good kinetic evidence to support the existence of either an acyl enzyme intermediate, or a discrete tetrahedral species during the chemical transformation, although evidence for their existence has been sought *via* partitioning experiments (*182*). The recent transient kinetic studies of *Latt et al.* (*183*) clearly show the formation and decay of an intermediate during the hydrolysis of fluorescent polypeptides (Fig. 12). Although the authors made no direct comment on the interpretation of this transient species, it would appear from inspection of Fig. 12 that its rate of formation could be an order of magnitude (or more) slower than the rates usually ascribed to diffusion-limited enzyme-substrate reactions. Hence, it is possible that the initial rapid transient represents a step or steps subsequent to the formation of the enzyme-substrate complex.

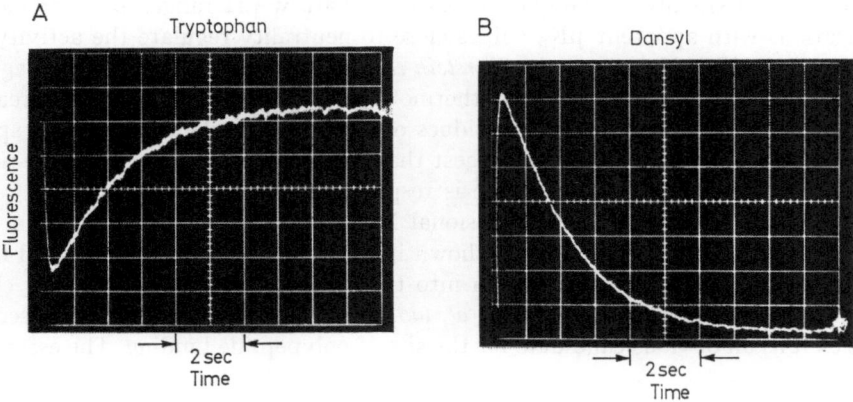

Fig. 12. Enzyme tryptophan (A) and substrate dansyl (B) fluorescence during the time course of zinc carboxypeptidase catalyzed hydrolysis of DNs-Gly-L-Phe. Equal volume solutions of substrate and of enzyme, both 2.5×10^{-4} M, in 1 M NaCl-0.02 M Tris, pH 7.5, 25°, were mixed and the fluorescence of either tryptophan (A) or dansyl (B) was measured as a function of time under stopped-flow conditions in parallel samples, as shown by the oscilloscope tracings. Excitation was at 285 nm. Scale sensitivities for (A) and (B) are 100 mV/div. The existence of the E · S complex is signalled either by (A) the *suppression* of enzyme tryptophan fluorescence (quenching by the dansyl group) or (B) *enhancement* of the substrate dansyl group fluorescence (energy transfer from enzyme tryptophan). Taken from Ref. (*182*) with permission

5. *Bacillus thermoproteolyticus* Thermolysin

Thermolysin is a potent endopeptidase isolated from the organism *Bacillus thermoproteolyticus*. In contrast to the other zinc metalloenzymes discussed in this review, thermolysin has not yet been subjected to rigorous chemical and kinetic investigation. Therefore, although the complete amino acid sequence is known (*184*), and although *Matthews* and co-workers (*43—46*) have carried the X-ray structural analysis to 2.3 Å resolution, it is neither possible to discuss the thermolysin catalytic mechanism nor to propose a role for zinc ion in thermolysin catalysis. Therefore, the following discussion is restricted to a brief review of the current status of the physical and chemical properties of thermolysin, and to a brief discussion of the 3-dimensional structure of thermolysin in the region of the essential zinc ion.

Thermolysin is specific for the hydrolysis of peptide linkages which occur adjacent to, but on the imino side of, a hydrophobic amino acid moiety (*i.e.*, X-Leu, X-Ileu, X-Phe, X-Val, where X is any other amino acid (*185—187*). Thus, the specificity of thermolytic cleavage is complementary to that of chymotryptic cleavage.

Thermolysin (MW 34,600) contains one zinc atom which has been shown by *Latt et al.* (*188*) to be essential for catalytic activity. *Feder* and *Schuck* (*187*) have made a preliminary investigation of the pH-dependence of k_{cat}/K_m for several N-*trans*-2-furylacryloyl dipeptides. The bell-shaped pH-dependence of this parameter indicates that the enzyme has a pH optimum near pH 7, and that the enzyme is maximally active over a relatively narrow pH range. Two (or more) ionizations with apparent pK_a values close to neutrality regulate the activity of the enzyme (*187*). More recently, *Burstein et al.* (*189*) have shown that the reagent ethoxyformic anhydride inhibits thermolysin. Although this reagent reacts nonspecifically with a variety of residues on the enzyme, the accumulated spectroscopic and chemical evidence suggest that the modification of a single histidyl imidazolyl moiety at the active site is responsible for the loss of activity.

A representation of the 3-dimensional X-ray structure of the enzyme in the vicinity of the essential zinc ion is shown in Fig. 13. The zinc ion is located in a deep cleft which divides the protein into two roughly equal portions. This cleft is open at each end, and *Colman et al.* (*45*) and *Matthews et al.* (*43, 44*) describe this cleft as an obvious candidate for the site of polypeptide binding. The essential

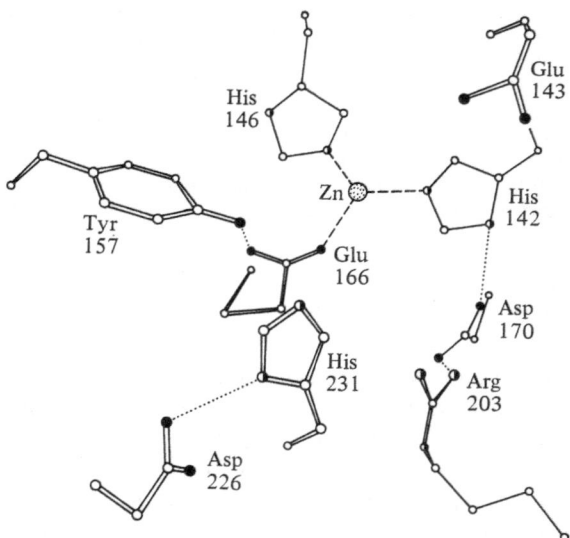

Fig. 13. Sketch illustrating the position of some of the residues in the active site of thermolysin. The direction of viewing is down the fourth ligand direction of the zinc. Taken from Ref. (*45*) with permission

zinc ion is four-coordinate. As shown in Fig. 13, the ligands derived from amino acid side chain residues are His-143, His-146, and Glu-166. The fourth ligand is believed to be a water molecule. These ligands are positioned in an approximately tetrahedral array about the metal ion with the water molecule projecting out into the cleft.

Several other notable features of this region of the enzyme surface can be seen in Fig. 13. The carboxyl group of Glu-143 is located in close proximity to the zinc ion in a configuration which bears a striking resemblance to the relative positions of Glu-270 and zinc ion in carboxypeptidase (see Fig. 11). Note the salt bridge between the guanidinium ion of Arg-203 and the carboxylate ion of Asp-170. And note also the close proximities of the imidazolyl moiety of His-231 ($\sim$5 Å) and the phenolic hydroxyl of Tyr-157 to the zinc ion. While some of these structural features are similar to those seen in the vicinity of the carboxypeptidase active site (i.e., the zinc ligands, the position of Glu-143, the proximities of Arg-203 and Tyr-157), Matthews et al. (44) argue that the differences are just as striking (i.e., the presence of His-231, and the salt bridge between Arg-203 and Asp-170). Preliminary comparisons of the 3-dimensional structures of the native enzyme and enzyme inhibitor complexes indicate that, in contrast to carboxypeptidase A, only small changes in the conformation of thermolysin accompany the binding of inhibitors (B. W. Matthews, personal communication).

Thus, although almost certainly some or all of the residues seen in Fig. 13 are of importance to the catalytic function of thermolysin, the available information is not sufficient to make possible a meaningful discussion of mechanism.

V. Conclusions — An Anthropomorphic View of Zinc Ion Catalysis

The small molecule and enzyme examples presented in this review offer considerable insight into the principles of catalysis which specify the roles played by zinc ion in homogeneous solution catalysis. It is perhaps worth emphasizing that zinc ion is capable of acting as a Lewis acid catalyst *only* if additional (weak) bonding forces between catalyst and substrate are sufficient to position the donor atom of the substrate within the first coordination sphere of the metal ion when the substrate donor atom is a comparatively weak ligand (viz., the carbonyl oxygens of an aldehyde or an amide). That is, Lewis acid catalysis can be significant for such a substrate only if the summation of bonding interactions between substrate and catalyst are sufficient to allow substrate to compete effectively with water for the metal ion inner coordination sphere.

The above principle is amply illustrated by the small molecule systems discussed in Section III. The roles proposed for zinc ion in the three enzyme systems discussed in Section IV also adhere to this principle. The accumulated experimental evidence makes it highly probable that zinc ion has a Lewis acid catalytic function both in the horse liver alcohol dehydrogenase-catalyzed reduction of aldehydes, and in the carboxypeptidase A-catalyzed hydrolysis of peptides. In contrast, the accumulated experimental evidence supports a role for zinc ion involving the enhancement of water nucleophilicity via inner sphere coordination in the carbonic anhydrase-catalyzed hydration of CO_2. The substrates for both LADH and carboxypeptidase are sufficiently large and chemically varied to conceivably provide the additional weak bonding interactions between site and substrate (*e.g.*, London dispersion forces, hydrogen bonds, charge-charge and charge-dipole forces) necessary to position the carbonyl oxygen of the substrate within the zinc ion inner coordination sphere. The size and the polarizability of CO_2 virtually exclude the possibility that there could occur additional weak bonding interactions between the site and CO_2 of sufficient strength to overcome the affinity of water for the inner coordination sphere. Therefore, it is not surprising that the carbonic anhydrase mechanism, in all probability, involves zinc ion in the activation of water for nucleophilic attack on CO_2.

While it makes good chemical sense in some instances for Zn(II)-coordinated H_2O to function catalytically as a nucleophile, it does not make good chemical sense for Zn(II)-coordinated H_2O to function as a general acid catalyst, since the extent of bond polarization brought about by the interaction of the aquo complex with substrate will be much less than that brought about by the direct interaction of the metal ion with substrate;

compare $Zn^{2+}\!-\!O\!-\!H\cdots S$ with $Zn^{2+}\cdots S$.
$$\underset{\displaystyle H}{\overset{\displaystyle |}{\phantom{Zn^{2+}-O-}}}$$

Indeed, the positioning of a water molecule between the metal ion and the substrate serves only to attenuate the Coulombic force field generated by the metal ion. Thus, since inner sphere coordination of substrate is a kinetically facile process for zinc aquo complexes, and since Lewis acid catalysis provides an inherently more energetic interaction, mechanisms which involve a zinc(II)-coordinated H_2O molecule as a general acid catalyst appear to be relatively unattractive alternatives to Lewis acid catalysis.

The above arguments provide a rational basis for discussion of the roles played by zinc ion in catalysis. Since zinc metalloenzymes are particularly abundant, it is legitimate to inquire as to the relative merits of Lewis acid catalysis by zinc ion *vis-á-vis* general acid-base catalysis. It is not possible to directly compare the catalytic effectiveness of Lewis acid catalysis to general acid catalysis, however, there are some notable differences between the two. Lewis acid catalysis differs from general acid-base catalysis in that the concentration of the effective species (assuming a $pK_a' \sim 9$ for the aquo ion) is independent of pH in the physiological pH range. In order that the effectiveness of a general acid catalyst show the same pH independence, it also must have a $pK_a' \sim 9$. Since the effectiveness of a general acid catalyst is inversely related to its pK_a', there is an obligatory trade-off between the relative effectiveness of the catalyst and the pH-dependent abundance of the reactive species (*190*).

If the comparison of Lewis acid catalysis *vs* general acid-base catalysis is extended to the specific examples of carboxypeptidase A and α-chymotrypsin, then it is interesting to note that the function of the carboxypeptidase active site catalytic residues (*33, 34*) appears to involve activation of the substrate for the chemical transformation primarily *via* Lewis acid catalysis (see Scheme III, Section IV). In contrast, the function of the α-chymotrypsin active site catalytic residues appears to involve the activation of the hydroxyl of Ser-195 for nucleophilic attack on the substrate primarily *via* general base catalysis (*165, 131—134*) as shown in Eq. (43). Thus, even though both enzymes carry out the same chemical transformation, the catalytic mechanisms of these two highly effective catalysts (to a first approximation) proceed *via* quite diverse pathways, one involving the activation of substrate, the other involving activation of a site catalytic residue.

Acknowledgments. The author gratefully acknowledges the support of a portion of the work described in this review by the National Science Foundation under Grant GB-31151. The author also thanks *Morton J. Gibian* for contributions in the form of theoretical discussions, and *William N. Libscomb, Carl-I. Brändén, Eila Zeppezauer, Poul Schack, Pier L. Luisi, James T. McFarland, Jean-F. Biellmann, K. K. Kannan,* and *John B. Neilands* for their criticisms and/or suggestions concerning the content of the manuscript.

VI. References

1. *Cotton, F. A., Wilkinson, G.:* Advanced inorganic chemistry, third edition, pp. 644—664 and 503—527. New York: John Wiley and Sons 1972.
2. *Eigen, M., Wilkins, R. G.:* In: Mechanisms of inorganic reactions (ed. by Gould, R. F.), pp. 55—80. Washington, D. C.: American Chemical Society 1964.
3. *Eigen, M.:* Pure Appl. Chem. 6, 105 (1963).
4. *Eigen, M., Hammes, G. G.:* Advan. Enzymol. 25, 1 (1964).
5. *Wilkins, R. G.:* Quart. Rev. (London) 16, 316 (1962).
6. *Lundberg, B. K. S.:* Acta Cryst. 21, 901 (1966).
7. *Belford, R. L., Chasteen, N. D., Hitchman, M. A., Hon, P.-K., Pfluger, C. E., Paul, I. C.:* Inorg. Chem. 8, 1312 (1969).
8. *Fratiello, A., Kubo, U., Deak, S., Sanchez, B., Schuster, R. E.:* Inorg. Chem. 10, 2552 (1971).
9. *Wilkins, R. G.:* Inorg. Chem. 3, 520 (1964).
10. *Margerun, D. W., Rosen, H. M.:* J. Am. Chem. Soc. 89, 1088 (1967).
11. *Kustin, K., Pasternack, R. F.:* J. Phys. Chem. 73, 1 (1969).
12. *Bernhard, S. A.:* The structure and function of enzymes. pp. 16, 75—76. New York: W. A. Benjamin, Inc. 1968.
13. *Zeltmann, A. H., Morgan, L. O.:* Inorg. Chem. 9, 2522 (1970).
14. *Sillen, L. G., Martell, A. E.:* Stability constants of metal-ion complexes, Parts II, III. London: The Chemical Society, Burlington House 1964, 1970.
15. *Bruice, T. C.:* In: The enzymes II, third edition, pp. 217—279. New York: Academic Press 1970.
16. *Page, M. I., Jencks, W. P.:* Proc. Natl. Acad. Sci. U. S. 66, 445 (1970).
17. *Dunn, M. F., Bernhard, S. A.:* In: Techniques of chemistry, Vol. VI, Part I: Investigations of rates and mechanisms of reactions (ed. by Lewis, E. S.), pp. 620—691. New York: Wiley-Interscience 1974.
18. *Storm, D. R., Koshland, D. E.:* Proc. Natl. Acad. Sci. U. S. 68, 658 (1971).
19. *Breslow, R., Fairweather, R., Keana, J.:* J. Am. Chem. Soc. 89, 2135 (1967).
20. *Creighton, D. J., Sigman, D. S.:* J. Am. Chem. Soc. 93, 6314 (1971).

21. *Shinkai, S., Bruice, T. C.:* J. Am. Chem. Soc. *94,* 8258 (1972).
22. *Shinkai, S., Bruice, T. C.:* Biochemistry *12,* 1750 (1973).
23. *Creighton, D. J., Hajdu, J., Mooser, G., Sigman, D. S.:* J. Am. Chem. Soc. *95,* 6855 (6973).
24. *Metzler, D. E., Snell, E. E.:* J. Am. Chem. Soc. *74,* 979 (1952).
25. *Breslow, R. B., Chipman, D.:* J. Am. Chem. Soc. *87,* 4195 (1965).
26. *Lloyd, G. J., Cooperman, B. S.:* J. Am. Chem. Soc. *93,* 4883 (1971).
27. *Sigman, D. S., Jorgensen, C. T.:* J. Am. Chem. Soc. *94,* 1724 (1972).
28. *Buckingham, D. A., Foster, D. M., Sargeson, A. M.:* J. Am. Chem. Soc. *91,* 3451, 4102 (1969).
29. *Kirkwood, J. G., Westheimer, F. H.:* J. Chem. Phys. *6,* 506 (1938).
30. *Coleman, J. E.:* Progr. Bioorg. Chem. *1,* 160—344 (1971).
31. *Lindskog, S., Henderson, L. E., Kannan, K. K., Liljas, A., Nyman, P. O., Strandberg, B.:* The enzymes V, (ed. by Boyer, P.), third edition, pp. 587—665. New York: Academic Press 1971.
32. *Lindskog, S.:* In: Structure and bonding (ed. by Hemmerich, P. *et al.*), Vol. 8, pp. 153—192. Berlin-Heidelberg-New York: Springer 1970.
33a. *Lipscomb, W. N., Hartsuck, A., Recke, G. N., Quiocho, F. A., Bethge, P. W., Ludwig, M. L., Steitz, T. A., Muirhead, H., Coppola, C.:* Brookhaven Symp. Biol. *21,* 24 (1968).
33b. *Hartsuck, J. A., Lipscomb, W. N.:* In: The enzymes III (ed. by Boyer, P.), third edition, pp. 1—56. New York: Academic Press 1971.
34. *Quiocho, F. A., Lipscomb, W. N.:* In: Advances in protein chemistry *25* (ed. by Anfinsen, C. B., Edsall, J. T., Richards, F. M.), pp. 1—78. New York: 1971.
35. *Sund, H., Theorell, H.:* In: The enzymes VII (ed. by Boyer, P.), second edition, pp. 1—83. New York: Academic Press 1963.
36. *Reeke, G. N., Hartsuck, J. A., Ludwig, M. L., Quiocho, F. A., Steitz, T. A., Lipscomb, W. N.:* Proc. Natl. Acad. Sci. U. S. *58,* 2220 (1967).
37. *Lipscomb, W. N., Hartsuck, J. A., Reeke, G. N., Quiocho, F. A., Bethge, P. H., Ludwig, M. L., Steitz, T. A., Muirhead, H., Cappola, J. C.:* Brookhaven Symp. Biol. *21,* 24 (1968).
38. *Lipscomb, W. N., Reeke, G. N., Hartsuck, J. A., Quiocho, F. A., Bethge, P. H.:* Phil. Trans. Roy. Soc. London *257,* 177 (1970).
39. *Quiocho, F. A., Bethge, P. H., Lipscomb, W. N., Studebaker, J. F., Brown, R. D., Koenig, S. H.:* Cold Spring Harbor Symp. Quant. Biol. *36,* 117 (1971).
40. *Kannan, K. K., Liljas, A., Waara, I., Bergstén, P.-C., Lovgren, S., Strandberg, B., Bergtsson, U., Carlbom, U., Fridborg, K., Järup, L., Petef, M.:* Cold Spring Harbor Symp. Quant. Bicl. *36,* 221 (1971).
41. *Bergstén, P.-C., Waara, I., Lövgen, S., Liljas, A., Kannan, K. K., Bengtsson, U.:* In: Alfred Benzen Symposium IV, pp. 363—383. Copenhagen: 1971.
42. *Liljas, A., Kannan, K. K., Bergstén, P.-C., Waara, I., Fridborg, K., Strandberg, B., Carlbom, U., Järup, L, Lövgren, S.:* Nature, New Biol. *235,* 131 (1972).
43. *Matthews, B. W., Jansonius, J. N., Colman, P. M., Schoenborn, B. P., Dupourque, D.:* Nature, New Biol. *238,* 37 (1972).
44. *Matthews, B. W., Colman, P. M., Jansonius, J. N., Titani, K., Walsh, K. A., Neurath, H.:* Nature, New Biol. *238,* 41 (1972).
45. *Colman, P. M., Jansonius, J. N., Matthews, B. W.:* J. Mol. Biol. *70,* 701 (1972).
46. *Matthews, B. W., Weaver, L. H.:* Biochemistry *13,* 1719 (1974).
47. *Brändén, C.-I., Eklund, H., Nordström, B., Boiwe, T., Söderlund, G., Zeppezauer, E., Ohlsson, I., Åkeson, Å.:* Proc. Natl. Acad. Sci. U. S. *70,* 2439 (1973). — *Eklund, H., Nordström, B., Zeppezauer, E., Söderlund, G., Ohlsson, I., Bowie, T., Brändén, C.-I.:* FEBS Letters *44,* 200 (1974).
48. *Blundell, T. L., Cutfield, J. F., Dodson, E. J., Dodson, G. G., Hodgkin, D. C., Mercola, D. A.:* Cold Spring Harbor Symp. Quant. Biol. *36,* 233 (1971).
49. *Adams, M. J., Blundell, T. L., Dodson, E. J., Dodson, G. G., Vijayan, M., Baker, E. N., Harding, M. M., Hodgkin, D. C., Rimmer, B., Sheats, S.:* Nature *224,* 491 (1969). — *Blundell, T. L., Dodson, G. G., Dodson, E. J., Hodgkin, D. C., Vijayan, M.:* Recent Progr. Hormone Res. *27,* 1 (1971).
50. *Reimann, J. E., Jencks, W. P.:* J. Am. Chem. Soc. *88,* 3973 (1966).

51. *Pietruszko, R., Ringold, H. J., Li, T. K., Vallee, B. L., Åkeson, Å., Theorell, H.:* Nature *221*, 440 (1969).
52. *Bjorkhem, I.:* European J. Biochem. *30*, 441 (1972).
53. *Anderson, B. M., Kaplan, N. O.:* J. Biol. Chem. *234*, 1226 (1959).
54. *Anderson, B. M., Ciotti, J. C., Kaplan, N. O.:* J. Biol. Chem. *234*, 1219 (1959).
55. *Biellmann, J.-F., Hirth, C. G., Jung, J. M., Roseheimer, N., Wrixon, A. D.:* European J. Biochem. *41*, 517 (1974).
56. *Jörnvall, H.:* Proc. Natl. Acad. Sci. U. S. *70*, 2295 (1973).
57. *Kagi, J. H. K., Vallee, B. L.:* J. Biol. Chem. *235*, 3188 (1960).
58. *Åkeson, Å.:* Biochem. Biophys. Res. Commun. *17*, 211 (1964).
59. *Oppenheimer, H. L., Green, R. W., McKay, R. H.:* Arch. Biochem. Biophys. *119*, 552 (1967).
60. *Drum, D. E., Li., T.-K., Vallee, B. L.:* Biochemistry *8*, 3792 (1969).
61. *Drum, D. E., Vallee, B. L.:* Biochemistry *9*, 4078 (1970).
62. *Iweibo, I., Weiner, H.:* Biochemistry *11*, 1003 (1972).
63. *Coleman, P. L., Iweibo, I., Weiner, H.:* Biochemistry *11*, 1010 (1972).
64. *Takahashi, M., Harvey, R. A.:* Biochemistry *12*, 4743 (1973).
65. *Theorell, H., McKinley-McKee, J. S.:* Acta Chem. Scand. *15*, 1811 (1961).
66. *Taniguchi, S., Theorell, H., wkeson, w.:* Acta Chem. Scand. *21*, 1903 (1967).
67. *Dalziel, K.:* J. Biol. Chem. *238*, 2850 (1963).
68. *Dunn, M. F.:* Biochemistry *13*, 1146 (1974).
69. *Everse, J.:* Mol. Pharmacol. *9*, 199 (1973).
70. *Winer, A. D., Theorell, H.:* Acta Chem. Scand. *13*, 1038 (1959); *14*, 1729 (1960).
71. *Theorell, H., Yonetani, T.:* Biochem. Z. *338*, 537 (1963).
72. *Dunn, M. F., Hutchison, J. S.:* Biochemistry *12*, 4882 (1973).
73. *Hansch, C., Schaeffer, J., Kerley, R.:* J. Biol. Chem. *247*, 4703 (1972).
74. *Sarma, R. H., Woronick, C. L.:* Biochemistry *11*, 170 (1972).
75. *Sigman, D. S.:* J. Biol. Chem. *242*, 3815 (1967).
76. *Theorell, H., Chance, B.:* Acta Chem. Scand. *5*, 1127 (1951).
77. *Wratten, C. C., Cleland, W. W.:* Biochemistry *2*, 935 (1963); *4*, 2442 (1965).
78. *Bernhard, S. A., Dunn, M. F., Luisi, P. L., Schack, P.:* Biochemistry 9, 185 (1970).
79. *Dunn, M. F., Bernhard, S. A.:* Biochemistry *10*, 4569 (1971).
80. *McFarland, J. T., Bernhard, S. A.:* Biochemistry *11*, 1486 (1972).
81. *Luisi, P. L., Favilla, R.:* Biochemistry *11*, 2303 (1972).
82. *Geraci, G., Gibson, Q. H.:* J. Biol. Chem. *242*, 4275 (1967).
83. *Shore, J. D.:* Biochemistry *8*, 1588 (1969).
84. *Shore, J. D., Gutfreund, H.:* Biochemistry *9*, 4655 (1970).
85. *Shore, J. D., Gutfreund, H., Santiago. D., Santiago, P.:* In: 1st Intern. Symp. Alcohol Aldehyde Metabol. Systems, Stockholm, July 9—11, 1973.
86. *Jacobs, J. W., McFarland, J. T., Wainer, I., Jeanmaier, D., Ham, C., Hamm, K., Wnuk, M., Lamm, M.:* Biochemistry *13*, 60 (1974).
87. *Matthews, B., Bernhard, S. A.:* Ann. Rev. Biophys. Bioeng. *2*, 257, (1973).
88. *Malhotra, O. P., Seydoux, F., Bernhard, S. A.:* CRC Critical Rev. Biochem. *1974*, 227.
89. *Dunn, M. F., Biellmann, J.-F., Branlant, G.:* Biochemistry, in press.
90. *Biellmann, J.-F., Jung, M. J.:* European J. Biochem. *19*, 130 (1971).
91. *Wallenfels, K., Sund, H.:* Biochem. Z. *329*, 59 (1957).
92. *Ables, R. H., Hutton, R. F., Westheimer, F.:* J. Am. Chem. Soc. *79*, 712 (1957).
93. *Kosower, E. M.:* Biochem. Biophys. Acta *56*, 474 (1962).
94. *Schellenberg, K.:* In: Intern. Symposium on Pyridine Nucleotide-Dependent Dehydrogenases (ed. by *Sund, H.*), pp. 15—29. Berlin—Heidelberg—New York: Springer 1970.
95. *Hamilton, G. A.:* Progr. Bioorg. Chem. *1*, 112—115, (1971).
96. *Bruice, T. C., Benkovic, S. J.:* Bioorganic mechanisms, Vol. II, pp. 343—349. New York: Benjamin 1966.
97. *Edsall, J. T.:* Harvey Lectures *62*, 191 (1968).
98. *Lin, K.-I., Deutsch, H. F.:* J. Biol. Chem. *248*, 1885 (1973).
99. *Henderson, L. E., Henriksson, D., Nyman, P. O.:* Biochem. Biophys. Res. Commun. *52*, 1388 (1973).

100. *Lindskog, S., Malmström, B. G.:* J. Biol. Chem. *237,* 1129 (1962).
101. *Pocker, Y., Meany, J. E.:* J. Am. Chem. Soc. *87,* 1809 (1965).
102. *Tashian, R. E., Plato, C. C., Shows, T. B.:* Science *140,* 53 (1963).
103. *Schneider, F., Liefländer, M.:* Z. Physiol. Chem. *334,* 279 (1963).
104. *Pocker, Y., Stone, J. T.:* J. Am. Chem. Soc. *87,* 5497 (1965).
105. *Kaiser, E. T., Lo, K.-W.:* J. Am. Chem. Soc. *91,* 4912 (1969).
106. *Waara, I., Lövgren, S., Liljas, A., Kannan, K. K., Bergstén, P.-C.:* In: Hemoglobin and red cell structure and function (ed. by Brewer, G. J.), pp. 169—187. New York: Plenum 1973.
107. *DeVoe, H., Kistiakowsky, G. B.:* J. Am. Chem. Soc. *83,* 274 (1961).
108. *Kernohan, J. C.:* Biochim. Biophys. Acta *81,* 346 (1964); *96,* 304 (1965); *118,* 405 (1966).
109. *Khalifah, R. G.:* J. Biol. Chem. *246,* 2561 (1971).
110. *Magid, E.:* Biochim. Biophys. Acta *151,* 236 (1968).
111. *Koenig, S. H., Brown, R. D.:* Proc. Natl. Acad. Sci. U. S. *69,* 2422 (1972).
112. *Khalifah, R. G.:* Proc. Natl. Acad. Sci. U. S. *70,* 1986 (1973).
113. *Lindskog, S., Coleman, J. E.:* Proc. Natl. Acad. Sci. U. S. *70,* 2505 (1973).
114. *Eigen, M.:* Angew. Chem. Intern. Ed. Engl. *3,* 1 (1964).
115. *Alberty, R. A., Hammes, G. G.:* J. Phys. Chem. *62,* 154 (1958).
116. *Taylor, P. W., Burgen, A. S. V.:* Biochemistry *10,* 3859 (1971).
117. *Taylor, P. W., Feeney, J., Burgen, A. S. V.:* Biochemistry *10,* 3866 (1971).
118. *Gerber, K., Ng, F. T. T., Pizer, R., Wilkins, R. G.:* Biochemistry *13,* 2663 (1974).
119. *Taylor, P. W., King, R. W., Burgen, A. S. V.:* Biochemistry *9,* 2638, 3894 (1970).
120. *King, R. W., Roberts, G. C. K.:* Biochemistry *10,* 558 (1971).
121. *Lindskog, S.:* J. Biol. Chem. *238,* 945 (1963); Biochemistry *5,* 2641 (1966).
122. *Feeney, J., Burgen, A. S. V.:* European J. Biochem. *34,* 107 (1973).
123. *Lanir, A., Navon, G.:* Biochemistry *11,* 3536 (1972).
124. *Chen, R. F., Kernohan, J. C.:* J. Biol. Chem. *242,* 5813 (1967).
125. *Riepe, M. E., Wang, J. H.:* J. Biol. Chem. *243,* 2779 (1968).
126. *Fabry, M. E. R., Koenig, S. H., Schillinger, W. E.:* J. Biol. Chem. *245,* 4256 (1970).
127. *Coleman, J. E., Coleman, R. V.:* J. Biol. Chem. *247,* 4718 (1972).
128. *Grell, E., Bray, R. C.:* Biochem. Biophys. Acta *236,* 503 (1971).
129. *Thorslund, A., Lindskog, S.:* European J. Biochem. *3,* 117 (1968). — *Lindskog, S., Thorslund, A.:* European J. Biochem. *3,* 453 (1968).
130. *Olander, J., Bosen, S. F., Kaiser, E. T.:* J. Am. Chem. Soc. *95,* 4473, 1616 (1973).
131. *Blow, D. M., Birktoft, J. J., Hartley, B. S.:* Nature *221,* 337 (1969).
132. *Hunkapiller, M. W., Smallcombe, S. H., Whitaker, D. R., Richards, J. H.:* Biochemistry *12,* 4732 (1973).
133. *Stroud, R. M.:* Sci. Am. *231,* 74 (1974). — *Krieger, M., Kay, L. M., Stroud, R. M.:* J. Mol. Biol. *83,* 209 (1974).
134. *Robillard, G., Shulman, R. G.:* J. Mol. Biol. *71,* 507 (1972).
135. *Gothe, P. O., Nyman, P. O.:* FEBS Letters *21,* 159 (1972).
136. *Henderson, R., Wright, C. S., Hess, G. P., Blow, D. M.:* Cold Spring Harbor Symp. Quant. Biol. *36,* 63 (1971). Also, *Nakagawa, Y., Bender, M. L.:* Biochemistry *9,* 259 (1970).
137. *Coleman, J. E.:* J. Biol. Chem. *242,* 5212 (1967).
138. *Davis, R. P.:* Enzymes *5,* 545 (1961).
139. *Oppenheimer, H. L., Labouesse, B., Hess, G. P.:* J. Biol. Chem. *241,* 2720 (1966).
140. *Hinz, H. J., Shiao, D. D. F., Sturtevant, J. M.:* Biochemistry *10,* 1347 (1971).
141. *Parsons, S. M., Raftery, M. A.:* Biochemistry *11,* 1623, 1630, 1633 (1972).
142. *Freisheim, J. H., Walsh, K. A., Neurath, H.:* Biochemistry *6,* 3010, 3020 (1967).
143. *Hanson, H. T., Smith, E. L.:* J. Biol. Chem. *175,* 833 (1948); *179,* 815 (1949).
144. *Abramowitz, N., Schechter, I., Berger, A.:* Biochem. Biophys. Res. Commun. *29,* 862 (1967).
145. *Riordan, J. F., Vallee, B. L.:* Biochemistry *2,* 1460 (1963).
146. *McClure, W. O., Neurath, H., Walsh, K. A.:* Biochemistry *3,* 1897 (1964).
147. *Bender, M. L., Whitaker, J. R., Menger, F.:* Proc. Natl. Acad. Sci. U. S. *53,* 711 (1965).
148. *Carson, F. W., Kaiser, E. T.:* J. Am. Chem. Soc. *88,* 1212 (1966).

149. *Vallee, B. L., Riordan, J. F., Bethune, J. L., Coombs, T. L., Auld, D. S.:* Biochemistry 7, 3547 (1968).
150. *Auld, D. S., Vallee, B. L.:* Biochemistry 9, 602 (1970).
151. *Auld, D. S., Vallee, B. L.:* Biochemistry 9, 4352 (1970).
152. *Davies, R. C., Riordan, J. F., Auld, D. S., Vallee, B. L.:* Biochemistry 7, 1090 (1968).
153. *Rupley, J. A., Gates, V.:* Proc. Natl. Acad. Sci. U. S. 57, 496 (1967).
154. *Chipman, D. M., Grisaro, V., Sharon, N.:* J. Biol. Chem. 242, 4388 (1967).
155. *Segal, D. M., Cohen, G. H., Davies, D. R., Powers, J. C., Wilcox, P. E.:* Cold Spring Harbor Symp. Quant. Biol. 36, 85 (1971).
156. *Rühlmann, A., Kukla, D., Schwager, P., Bartels, K., Huber, R.:* J. Mol. Biol. 77, 417 (1973).
157. *Shotton, D. M., White, N. J., Watson, H. C.:* Cold Spring Harbor Symp. Quant. Biol. 36, 91 (1971).
158. *Kraut, J., Robertus, J. D., Birktoft, J. J., Alden, R. A.:* Cold Spring Harbor Symp. Quant. Biol. 36, 117 (1971).
159. *Ingram, J. M., Wood, W. A.:* J. Biol. Chem. 241, 3256 (1966).
160. *Bernhard, S. A.:* J. Cellular Comp. Physiol. 54, (Suppl. 1), 256 (1959).
161. *Hein, G. E., Niemann, C.:* J. Am. Chem. Soc. 84, 4495 (1962).
162. *Koshland, Jr., D. E.:* Proc. Natl. Acad. Sci. U. S. 44, 98 (1958).
163. *Pauling, L.:* Am. Scientist 36, 51 (1948); Nature 161, 707 (1948).
164. *French, T. C., Yu, N. G., Auld, D. S.:* Biochemistry 13, 2877 (1974).
165. *Zerner, B., Bender, M. L.:* J. Am. Chem. Soc. 83, 2267 (1961); 85, 356 (1963). — *Bender, M. L., Clement, G. E., Kezdy, F. J., Zerner, B.:* J. Am. Chem. Soc. 85, 358 (1963).
166. *Latt, S. A., Vallee, B. L.:* Biochemistry 10, 4263 (1971).
167. *Sokolovsky, M., Vallee, B. L.:* Biochemistry 6, 700 (1967).
168. *Riordan, J. F., Sokolovsky, M. Vallee, B. L.:* Biochemistry 6, 358, (1967).
169. *Riordan, J. F., Sokolovsky, M., Vallee, B. L.:* Biochemistry 6, 3609 (1967).
170. *Coleman, J. E., Pulido, P., Vallee, B. L.:* Biochemistry 5, 2019 (1966).
171. *Kang, E. P., Storm, C. B.:* Biochem. Biophys. Res. Commun. 49, 621 (1972).
172. *Coleman, J. E., Vallee, B. L.:* J. Biol. Chem. 236, 2244 (1961).
173. *Zisapel, N., Sokolovsky, M.:* Biochem. Biophys. Res. Commun. 53, 722 (1973).
174. *Navon, G., Shulman, R. G., Wyluda, B. J., Yamane, T.:* Proc. Natl. Acad. Sci. 60, 86 (1968); J. Mol. Biol. 51, 15 (1970).
175. *Koenig, S. H., Brown, R. D., Studebaker, J.:* Cold Spring Harbor Symp. Quant. Biol. 36, 551 (1971).
176. *Johansen, J. T., Vallee, B. L.:* Proc. Natl. Acad. Sci. U. S. 70, 2006 (1973); 68, 2532 (1971).
177. *Quiocho, F. A., McMurray, C. H., Lipscomb, W. N.:* Proc. Natl. Acad. Sci. U. S. 69, 2850 (1972).
178. *Hass, G. M., Neurath, H.:* Biochemistry 10, 3535, 3541 (1971).
179. *Pétra, P. H.:* Biochemistry 10, 3163 (1971).
180. *Pétra, P. H., Neurath, H.:* Biochemistry 10, 3171 (1971).
181. *Lipscomb, W. N.:* Tetrahedron 30, 1725 (1974).
182. *Hall, P. L., Kaiser, E. T.:* Biochem. Biophys. Res. Commun. 29, 205 (1967).
183. *Latt, S. A., Auld, D. S., Vallee, B. L.:* Biochemistry 11, 3015 (1972).
184. *Titani, K., Hermodson, M. A., Ericsson, L. H., Walsh, K. A., Neurath, H.:* Nature New Biol. 238, 35 (1972).
185. *Bradshaw, R. A.:* Biochemistry 8, 3871 (1969).
186. *Matsubara, H., Singer, A., Sasaki, R., Jukes, T. H.:* Arch. Biochem. Biophys. 115, 324 (1966).
187. *Feder, J., Schuck, J. M.:* Biochemistry 9, 2784 (1970).
188. *Latt, S. A., Holmquist, B., Vallee, B. L.:* Biochem. Biophys. Res. Commun. 37, 333 (1969).
189. *Burstein, Y., Walsh, K. A., Neurath, H.:* Biochemistry 13, 205 (1974).
190. *Brønsted, J. N., Pederson, K. J.:* Z. Physik. Chem. (Leipzig) 108, 185 (1923).

Received August 12, 1974

Kinetics and Mechanism of Metalloporphyrin Formation

Walter Schneider

Laboratorium für Anorganische Chemie der ETH, 8006 Zürich, Switzerland

Table of Contents

1. Introduction

From the point of view of coordination chemistry, ferric haems, and chlorophylls are spectacular cases of very stable complexes that are formed at much slower rates than the usual complexes of iron and magnesium. *Falk (1)* has summarized the earlier work done on the kinetics of metalloporphyrin formation when there was little information available on the behaviour of different metal ions under comparable conditions. Nevertheless, the scattered studies showed that the rates depend on the metal ion, the substituents at the porphin nucleus (I), the solvent, and the ligands of the metal ion. The last point was made for the first time by *Lowe* and *Phillips (2)* who showed that bidentate ligands could induce a spread of rate constants over three orders of magnitude in the reaction of Cu(II) with mesoporphyrindimethylester solubilized by sodium dodecyl sulphate in aqueous solution. Much subsequent work has been devoted to the evaluation of rate laws yet without any clear picture of the mechanisms emerging, as outlined by *Hambright* (3) in a review on various aspects of the coordination chemistry of metalloporphyrins. Recent studies have emphasized the variation of ligands in the reaction of complexes of Cu(I), Cu(II), Fe(II), Fe(III) and Ni(II) with one particular synthetic porphyrin (4—9). A variety of porphyrins related to biologically relevant material was selected for kinetic studies by *Berezin* (10), whereas *Hambright* (11) preferred synthetic compounds suitable for aqueous solutions. The present selective review aims to reveal the nature of the dominant effects upon rates of formation of metalloporphyrins in well-defined solutions. Some clear conclusions can now be drawn with regard to the mechanisms involved. Various aspects of metalloporphyrin chemistry reviewed recently are: structures (12), redox reactions (13) and syntheses (14). Much effort has been devoted to spectra and electronic structure, particularly by *Gouterman* and his group (15). From all these areas, the structural data are most important in considering kinetic data at the present stage.

2. General Considerations[1])

2.1. Metal Incorporation Versus Ligand Substitution

Formally, two protons are replaced by a metal ion when a metalloporphyrin (MP) is formed according to the equation

$$M(z) + H_2P \longrightarrow MP^{z-2} + 2 H^+ \tag{1}$$

where z is the oxidation number, and the notation H_2P indicates the number of protons removable from the pyrrole nitrogens in the porphyrin base. In all cases to be considered, $M(z)$ refers to mononuclear complexes $M\Lambda^{(z+\lambda)}$, where Λ^λ stands for the entire first coordination sphere, involving equal or different ligands. The common term "incorporation" is an allusion to the entry of $M(z)$ into the center of the macrocyclic ligand P^{2-}, whereby the shell Λ is stripped off rather completely. It is by no means certain that the macrocyclic ligand H_2P could react with hydrated ions $M(H_2O)_n^{z+}$ by a stepwise substitution of water, in contrast to the behaviour of open-chain multidentate ligands. It has been pointed out that the porphyrin structure has the greatest tendency to force a mechanism of simultaneous multiple desolvation of the metal ion (16). Obviously, the latter requirement would be quite stringent if the porphyrin framework were virtually inflexible with respect to loss of planar conformation. At any rate, it is important to elucidate the

---

[1]) Notation used in this article: H_iP^{i-2}: porphyrin with $i = 1$, 2, 3, 4 protonated pyrrole nitrogens; net charge $(i - 2)$ of the central part irrespective of additional charges of substituents at the porphyrin skeleton (I). Code designations for specific porphyrins and metalloporphyrins are defined in the text. Net charges are indicated to the right of the code, if pertinent, within the text.

relationship between the mechanisms of normal substitution reactions on one hand and the substitution of Λ by H_2P on the other hand.

2.2. Relevant Properties of Porphyrin Ligands

With regard to kinetic studies in solution, particular attention must be given to

(i) association phenomena involving either porphyrin molecules only, or porphyrins and other solute species;

(ii) protonation of H_2P to the so-called diacids H_4P^{2+}, and deprotonation to the dianions P^{2-};

(iii) the flexibility of the porphyrin skeleton (I).

Fleischer (12) has reviewed evidence for nearly invariant bond distances and bond angles in the planar skeleton (I) of a variety of native and synthetic porphyrins containing substituents either in positions 1 to 8, or in the meso positions (α, β, γ, δ). Irrespective of substituents, porphyrins are soluble in aqueous solutions of mineral acids (0.1 to 1 M) thanks to the stepwise protonation of pyrrole nitrogens to monoacids H_3P^+ and diacids H_4P^{2+}, causing pronounced changes in the absorption spectra *(1, 17)*. However, the free bases H_2P of native porphyrins are soluble only in organic solvents. Similarly, meso-substituted porphyrins are insoluble in water unless peripheral charges are introduced. Examples of the latter type include TMPyP (II), TPyP (III), TPPS$_3$ (V), TPPS$_4$ (VI) and TCPP (VII). π interactions between porphyrins are quite pronounced in native porphyrins *(18)*, as indicated by association in nonaqueous media. In water and ethyleneglycol-water mixtures, TMPyP is monomeric up to about 10^{-4} M porphyrin, as is the case for TPyP at pH 0 and at pH 7.5, and for TCPP and TPPS$_3$ in ethyleneglycol (80% v/v)-water mixtures around pH 7 *(17)*. In aqueous solution, however, TCCP and TPPS$_3$ dimerize at pH 7.5. The stability constants of the dimers $(H_2P)_2$ have been determined by *Pasternack (17)* who also measured their rates of for-

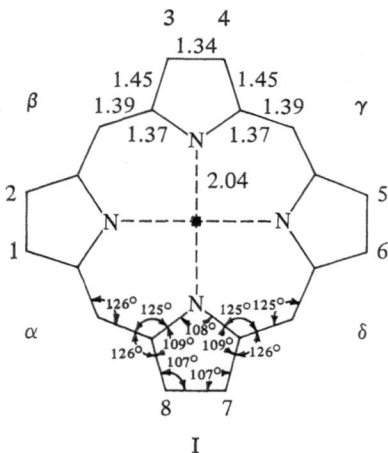

I

"Best" set of parameters for porphyrin skeleton from Ref. *(12)*

mation (Table 1). On the other hand, no aggregation of TPPS$_4$ has been observed at pH $\geq$ 3.0 (*19*). *Stein* and *Plane* (*20*) have prepared TEP (VIII) which provides the most stable of the dimers listed in Table 1. The activation energy $E_A \sim 12$ kcal/mole for dissociation indicates pronounced interaction in the dimer.

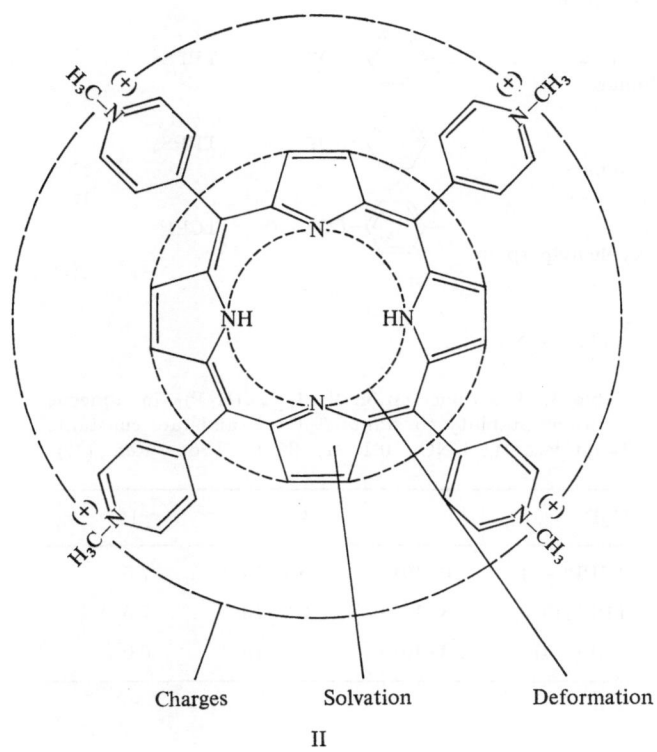

Charges   Solvation   Deformation

II

TMPyP

Tetra(N-methyl-pyridyl)-porphin

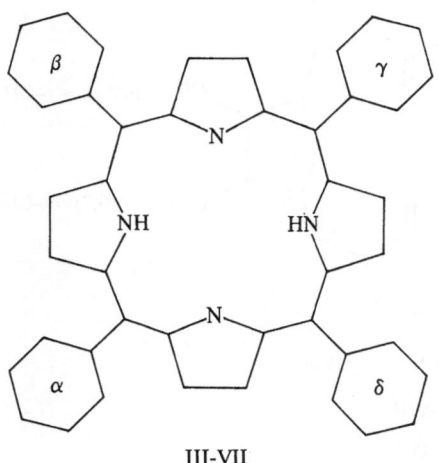

III-VII

III α, β, γ, δ =
Tetrapyridylporphin

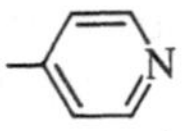

TPyP

IV α, β, γ, δ =
Tetraphenylporphin

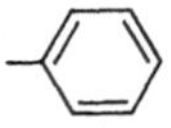

TPP

V α = H; β, γ, δ =
TPP-trisulfonate

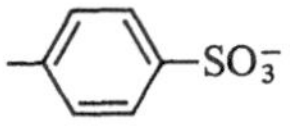

$-SO_3^-$

TPPS$_3$

VI α, β, γ, δ =
TPP-tetrasulfonate

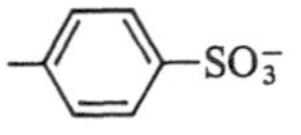

$-SO_3^-$

TPPS$_4$

VII α, β, γ, δ =
Tetracarboxyphenylporphin

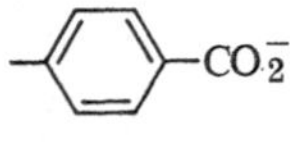

$-CO_2^-$

TCPP

Table 1. The dimerization $2\ H_2P \rightleftarrows (H_2P)_2$ in aqueous solution: Stability constants K (M^{-1}) and rate constants k_2 ($M^{-1}sec^{-1}$). KNO$_3$ 0.1 M; 25 °C. From Ref. (17)

H$_2$P	K	k_2	pH
TCPP(-4)	$4.6 \cdot 10^{+4}$	$6.4 \cdot 10^{+7}$	7.5
TPPS$_3$(-3)	$4.8 \cdot 10^{+4}$	$2.2 \cdot 10^{+8}$	7.5
TEP(>4)	$4.3 \cdot 10^{+6}$	$2.0 \cdot 10^{+8}$	6.0

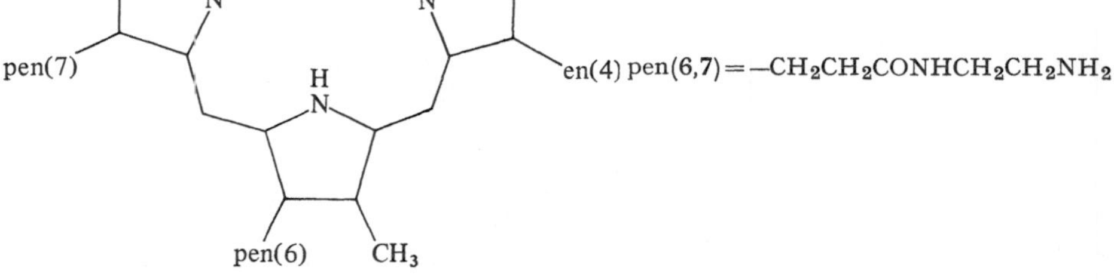

CH$_3$ en(2)

CH$_3$

CH$_3$ en(2,4) = $-CH_2CH_2NHCH_2CH_2NH_2$

pen(7)

en(4) pen(6,7) = $-CH_2CH_2CONHCH_2CH_2NH_2$

pen(6) CH$_3$

VIII

TEP

127

Although hydration of the N-methylpyridyl groups as well as steric and charge effects prevent the self-association of TMPyP($+4$) in aqueous solution, in association with smaller molecules such as para-toluenesulphonate it exhibits differentiated solvation behaviour, as discussed earlier by *Mauzerall* (*21*). Solutions used in the study of metal insertion usually contain buffer components and/or other ligands in addition to the solvent. In systems of this type, the composition of the domains close to the porphyrin may differ considerably from the bulk composition of the solution. Formula II indicates the different parts of the TMPyP molecule which are particularly involved in phenomena (i)—(iii).

It has been established by crystallographic data (*12*) and by H-NMR studies that the inner pyrrole hydrogens are opposite to each other. At room temperature the switch to equivalent positions is fast on the H-NMR time scale, as shown for several porphyrins dissolved in $CDCl_3/CS_2$ (*22, 23*). It is not known whether proton jumps are coupled to out-of-plane vibrations of the porphyrin skeleton. Double protonation to H_4P^{2+} freezes in an out-of-plane conformation of tetraphenylporphin as shown in Fig. 1, which is taken from Ref. (*12*). Another type of distortion has been postulated recently in connection with an intermediate observed in the reaction of $PtCl_4^{2-}$ with haematoporphyrin IX (*24*). It has been argued

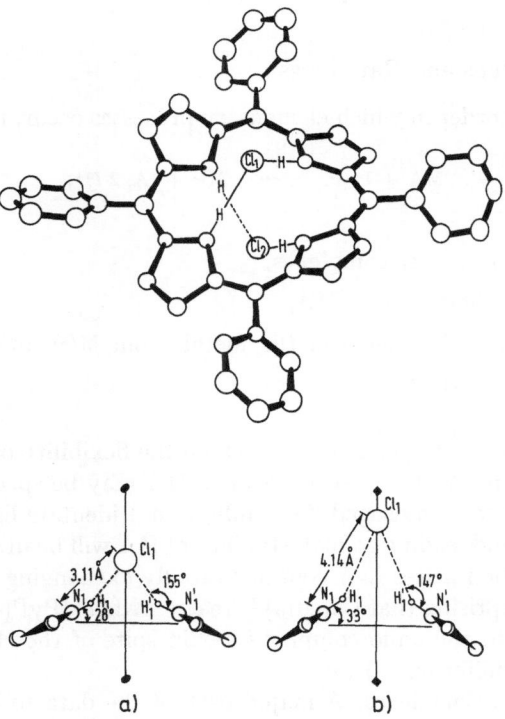

Fig. 1. The tilted porphin skeleton in the diacid of tetraphenylporphine (TPP); (*a*) and (*b*) represent diacids of TPP and of TPyP(III), respectively. Reproduced with permission from Ref. (*12*)

that the intermediate $PtCl_2(H_2P)$ contains the local unit $cis\text{-}PtCl_2N_2$, which involves nitrogens from two neighbouring pyrrole rings, a distortion in which the nitrogens are displaced pairwise in positions above and below the original average plane of the skeleton (I).

The diacids of TMPyP and $TPPS_4$ have closely spaced $pK_{3,4}$ in the pH intervals 1 to 2, and 4.5 to 5.5, respectively. Native porphyrins such as haematoporphyrin or uroporphyrin provide more complicated cases of subsequent protonation equilibria. Haematoporphyrin IX is insoluble in neutral aqueous solution but dissolves in dilute mineral acids, whereas its sodium salt Na_4P precipitates from NaOH 33% (w/v) (25). Hence, subsequent pK values cannot be determined in standard aqueous media. *Falk* (26) has listed pK_3 values determined in aqueous detergent solutions. Accepting $pK_3 \approx 5.5$ for deuteroporphyrindimethylester, it is safe to conclude that the series of four (two pyrrole N, two $-CO_2H$) pK values are rather narrowly spaced in the intervals from 2 to 6. Aggregation is overcome by adding 10% (v/v) pyridine to water, which provides monomeric $H_2P(-2)$ at $pH \geq 8$ (27). In the absence of pyridine, uroporphyrin I is monomeric in aqueous solution (27) but is insoluble below pH 4. Due to the eight carboxylic groups, the charge limit towards alkaline solutions is rather high in $H_2P(-8)$. Hence, a well-defined peripheral charge can only be secured in alkaline solutions at $pH \geq 9.5$ (11). Except for TMPyP, $pK_2 \sim 13$, no reliable values for pK_2 or pK_1 are available since they go even beyond 14 (28).

2.3. Conceptual Steps and Rate Laws

Irrespective of the order in which elementary processes occur, the overall reaction

$$M\Lambda + H_2P \longrightarrow MP + (\Lambda, 2\,H^+) \qquad (2)$$

involves

(i) formation of encounter complexes,

(ii) dissociation of ligands from $M\Lambda$,

(iii) attractive interaction between the metal atom $M(z)$ and pyrrole nitrogen,

(iv) deprotonation of H_2P.

Of course, stage (iii) depends very much on the flexibility of the porphyrin, as discussed in Section 2.2. To a certain degree, $M\Lambda$ may be specified right away if optimum pathways are envisaged. Certainly, a multidentate ligand occupying all sites at the metal and leading to high stability of $M\Lambda$ will be utterly unfavourable. A far better situation arises if Λ contains rapidly exchanging solvent molecules. Hence, it is not surprising that $Cu(edta)^{2-}$ reacts with $TMPyP(+4)$ at a negligible rate as compared to the aquo complex Cu^{2+} in spite of the charge $(2-)$, which favours the pre-equilibrium (i) (5).

2.3.1. Encounter Complexes. A major part of the data to be discussed refers to water soluble porphyrins with peripheral charges. In TMPyP and $TPPS_4$ the charged groups are unable to form inner-sphere complexes with $M(z)$. On the other hand, TPyP and uroporphyrin carry pyridyl and carboxylic groups, respectively,

129

i.e. ligand groups equivalent to pyridine and acetate. Hence, the latter two porphyrins exhibit rather high local concentrations of ligand groups.

Outer-sphere complexes have been discussed by several authors in the context of ligand substitution reactions of metal complexes (*29, 30*). We consider here an example relevant to reaction (2), the outer-sphere association of TMPyP($+4$) with $M\Lambda^x$, $x = 0$, ± 1, ± 2.

$$M\Lambda^x + H_2P^y \; \rightleftharpoons \; (M\Lambda, H_2P)^{x+y} \tag{3}$$

$$K_{0s} = \frac{[x+y]}{[x]\,[y]} = K_{0s}^0 \cdot \exp\left(-\frac{x \cdot y \cdot e_0^2}{\varepsilon\,k\,T \cdot r}\right) \cdot \pi_f \tag{4}$$

$$K_{0s}^0 = \frac{4\,\pi\,N \cdot a^3}{3000} \tag{5}$$

K_{0s} stability constant, molar scale

$[i]$ concentration of species in reaction (3) with charge i

ε dielectric constant

π_f activity factor

a radius sum of spherical particles

r distance between charges of contacting ions

Association constants K_{0s} have been estimated by using Eq. (4). introduced by *Fuoss* (*31*). It should be pointed out here that the Fuoss theory assumes spherical ions in contact with each other at a distance a (center to center). In these cases, r in Eq. (4) is identical with a in Eq. (5), which accounts for the association of uncharged molecules resulting from random collisions. It is doubtful whether Eqs. (4) and (5) can be applied to nonspherical ions such as TMPyP($+4$) containing peripheral charges. Moreover, solvent interactions including effects like those described in Section 2.2 can hardly be taken into account. However, the second and third factors to the right in Eq. (4) provide some indications about the effects on K_{0s} of charges x, y and of the ionic strength I. Assuming $M\Lambda^x = M(H_2O)_6^{2+}$ with radius 3.6 Å, and taking into account the structural parameters of TMPyP together with standard values for van der Waals parameters, an average value of $<a^3> = 1.2 \cdot 10^{+3}$ Å^3 has been evaluated (*7*). The average $<a>$, *i.e.* center of H_2P to center of $M(H_2O)_6^{2+}$, equals 10.4 Å, very close to $<a^3>^{1/3} = 10.7$ Å. Inserting $<a^3>$ into Eq. (5) provides a value

$$K_{0s}^0 \approx 3 \text{ M}^{-1}, \tag{6}$$

which is an order of a magnitude larger than the lowest value calculated for simple hydrated ions such as Na$^+$ and ClO$_4^-$ (*32*). Similarly, average values $<r> = 12.5$ Å, and $<r^{-1}> = 0.089 \approx <r>^{-1}$ have been calculated, where r is the distance between methylated pyridyl N and the center of $M\Lambda$. Hence,

$<r^{-1}> \neq <a^{-1}>$ has been inserted into the exponential in Eq. (4). Finally, the activity term π_t was taken to be

$$\log \pi_t = \frac{x \cdot y \cdot 2 A \cdot \sqrt{I}}{1 + B \cdot \alpha \sqrt{I}}. \qquad (7)$$

In Eq. (7), A and B are standard parameters of the Debye-Hückel theory (33) which associates α with the distance of closest approach. Remembering that the denominator in Eq. (7) takes care of short-range interactions, α could be identified with $<r> = 12.5$. On the other hand, the average distance between ions in 1:1 solutions of electrolytes approaches the distances between charges in TMPyP at concentrations of about 0.1 M[2]).

Therefore, short-range interactions actually refer to local groups, allowing distances as short as ~ 7 Å to the hydrated metal ion. The numbers in Fig. 2

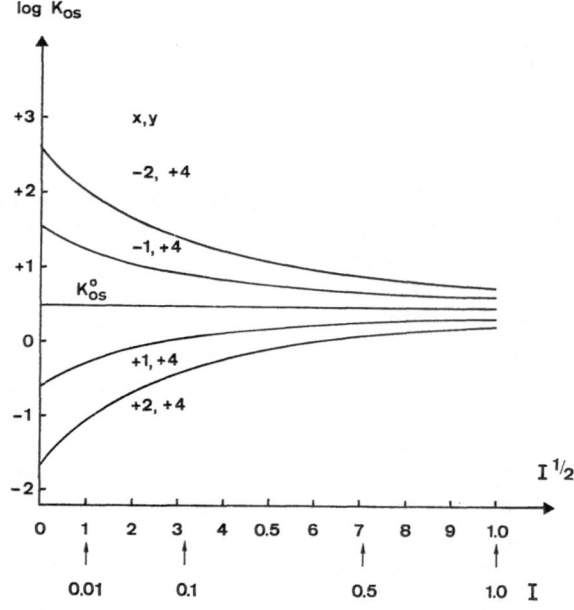

Fig. 2. Outer-sphere association of spherical ions (charge x) and non-spherical ions (charge y). Association constants K_{os} derived from Eq. (4,5,7). Ionic strength I. See text for assumptions about size and charge distribution

---

[2]) For 0.1 and 1 M solutions, average separations of ions are 20 Å and 9.4 Å, respectively. Positive charges in TMPyP are at distances 16 Å and 11 Å from each other.

are based on $\alpha = 10$ Å. Although these numerical values of K_{0s} cannot be reliable, they are significant as they indicate that

(i) electrostatic effects of long-range type are considerable at very low ionic strength; *e.g.* TMPyP($+4$) and TPPS$_4$(-4) should differ in K_{0s} for association with M_{aq}^{2+} by a factor as high as ca. 10^4;

(ii) towards higher ionic strength, $0.1 < I < 1.0$, as is typical for most solution studies, the spread for charge types $(+4, +2)$ vs. $(-4, +2)$ is much reduced, *i.e.* K_{0s} probably approaches K_{0s}^0.

At present, no data are available on the reactions of Zn^{2+} with the two porphyrins TMPyP and TPPS$_4$ under strictly comparable conditions. However, there are no major changes in the rate constants in Table 2, indicating that charge effects are screened by rather high ionic strengths according to conclusion (ii) above. The factor of about 3 by which k_2 (d) differs from k_2 (a) is due primarily to the rise in pH, as indicated by k_2 (b) $> k_2$ (a). However, the rates in case (b) quite clearly depend on the buffer concentration at fixed pH. Whereas a linear dependence of k_2 on the concentration of nitrate has been observed in cases (a) and (b), k_2 (c) remains virtually constant from 0.02 to 0.6 M NO_3^-, with I varying from 0.03 to 0.63. In these cases, ionic strength does not operate in the fashion indicated by Fig. 2. Therefore, specific ion effects are likely to be more important than general salt effects.

Table 2. Second-order rate constants k_2(M^{-1}sec^{-1}) for the reactions
$Zn^{2+} + H_2P^{4\pm} \rightarrow ZnP^{4\pm} + 2\,H^+$ (25 °C)
$Cu^{2+} + ZnP^{4\pm} \rightarrow CuP^{4\pm} + Zn^{2+}$ (40 °C)
$H_2P^{4+} = $ TMPyP(II); $H_2P^{4-} = $ TPPS$_4$(VI)

M_{aq}^{2+}	H_2P / MP	Medium			k_2	Ref.
a) Zn^{2+}	TMPyP^{4+}	NaNO$_3$	1.0 M;	pH 4.0	0.050	(*11*)
b)		NaNO$_3$	1.0 M;	pH 6.3	0.52	(*11*)
c) Zn^{2+}	TPPS$_4^{4-}$	NaNO$_3$	0.2 M;	pH 6.3	0.85	(*11*)
d)		NaClO$_4$	1.0 M;	pH 7.0	0.15	(*34*)
e) Cu^{2+}	Zn(TMPyP)$^{4+}$	NaNO$_3$	2.0 M;	pH 3.0	$1.3 \cdot 10^{-3}$	(*35*)
f) Cu^{2+}	Zn(TPPS$_4$)$^{4-}$	NaClO$_4$	0.5 M;	pH~4.0	$10 \cdot 10^{-3}$	(*34*)

a) Unbuffered solutions; $k_2 = k_3 \cdot [NO_3^-]$.
b) Buffer HB$^+$, B ($= 2,6$-lutidine); $k_2 = k_3 \cdot [NO_3^-]$.
c) PIPES buffer; base: piperazin-N,N'-bis(2-ethanesulfonate).
d) TRIS buffer.
e) } Unbuffered.
f) }

Turning to porphyrins with peripheral carboxylic groups, Kassner and Wang's data on the insertion of Fe(II) into haemato-, copro- and uroporphyrin deserve attention (*27*). Under strictly comparable conditions the rate constants increase by more than two orders of magnitude at low ionic strength.

Table 3. Kinetic data for the formation of Fe(II)-porphyrins in water-pyridine (10% v/v). Second-order rate constants k_2(M^{-1} sec^{-1}) and activation energies E_A (kcal $\cdot$ mole^{-1}); pH 8.2; I $\leq$ 0.001. From Ref. (27)

Porphyrin[b])	k_2 (25 °C)	E_A (± 1.0)
Haematoporphyrin IX (-2)[a])	0.22	24
Coproporphyrin I (-4)	8.2	24
Uroporphyrin I (-8)	84.0	25

[a]) Net charges in brackets refer to all carboxylic groups deprotonated. Incomplete deprotonation of uroporphyrin cannot be excluded.

[b]) Substituents in positions 1 to 8:
Haemato-: $CH_2CH_2CO_2H$(6,7); CH_3(1,3,5,8); $CHOHCH_3$(2,4)
Copro-: $CH_2CH_2CO_2H$(2,4,6,8); CH_3(1,3,5,7)
Uro- : $CH_2CH_2CO_2H$(2,4,6,8); CH_2CO_2H(1,3,5,7).

It may seem obvious that the effect is electrostatic in origin, like the large gap in log K_{0s} to the left of Fig. 2. However, the local concentration of carboxylic groups in uroporphyrin is of the order of 1 M, and coordination to Fe(II) cannot be ruled out. There is evidence for coordination of Cu(II) to peripheral amine groups in the reaction with TEP (VIII). It was found that rates did not depend linearly on Cu(II) when all other variables were kept constant (36).

2.3.2. Ligand Substitution in MΛ by the Porphyrin. Even in the most labile hydrated transition-metal ion, Cu^{2+}, ligand substitution generally occurs more slowly than the preceding diffusion step (37). It is useful to look at reaction (8) in relation to the much faster reactions (9) and (10), which have been carefully studied by Diebler (38).

$$Cu^{2+} + H_2P \longrightarrow CuP + 2\,H^+ \qquad \begin{matrix} k_2{}^{3)} \\ 0.6 \end{matrix} \qquad (8)$$

$$Cu^{2+} + dip \longrightarrow Cudip^{2+} \qquad 4 \cdot 10^{+7} \qquad (9)$$

$$Cu^{2+} + Hdip^+ \longrightarrow Cudip^{2+} + H^+ \qquad 3 \cdot 10^{+5} \qquad (10)$$

The Scheme

$$Cu^{2+}_{aq} + H_iL \underset{k_-}{\overset{k_+}{\rightleftharpoons}} Cu(H_2O)\,H_iL \overset{k_z}{\longrightarrow} CuL + i\,H^+ \qquad (11)$$

and

$$k_2 = \frac{k_+}{k_-}\,k_z = K_0\,k_z \qquad (12)$$

---

3) Second-order rate constants (M^{-1}sec^{-1}) for 25 °C, $I = 0.3$. $H_2P = TMPyP(+4)$.

can be adopted for our purpose, because the pre-equilibrium steps are well separated from the much slower outer sphere-inner sphere transformations. The intermediate species preceding the last step, $H_2O \rightarrow N$, is rather well defined for $L = dip$ but far less so for $L = H_2P$. It may be defined for the present purpose by placing hydrated Cu^{2+} in contact with the central part of H_2P, as indicated in Fig. 3.

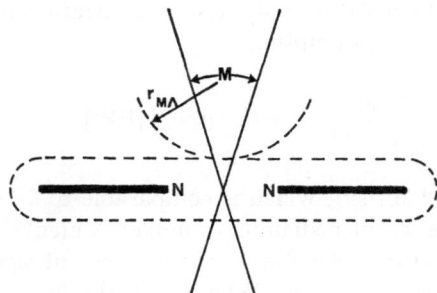

Fig. 3. Configuration of outer-sphere complex ($M\Lambda$, H_2P) preceding the insertion of the metal. Space angle indicated about $4\pi/50$

Consequently, K_0 refers to particular configurations of encounter complexes with assigned equilibrium constant K_{0s} (see Section 2.3.1). For small ligands, K_0 has frequently been approximated by K_{0s} as estimated by the Fuoss theory. However, if $L = H_2P$, it would be more appropriate to write

$$K_0 = \varkappa \cdot K_{0s}. \tag{13}$$

If the same statistical weight is given to all configurations of encounter complexes ($M\Lambda$, H_2P), $\varkappa$ may be as low as 10^{-2}. On the other hand, if the positions indicated in Fig. 3 were favoured by particular site interactions, $\varkappa$ could increase beyond one. Viewing at an optimum path, replacement of water by different ligands could contribute to faster insertion of a metal by increasing $\varkappa$. [The fact that dipyridyl complexes of Cu^{2+} and Ni^{2+} are formed more slowly than corresponding acetato complexes may be due, in part, to $\varkappa(ac^-) > \varkappa(dip)$].

In Eq. (11), not only the pre-equilibrium steps, but also the proper insertion steps are of a more complex nature in the case of porphyrin. The rate parameter k_z refers to an overall process which may have further built-in pre-equilibrium-type situations.

It is an important fact that reaction (10) is slower than (9) by two orders of magnitude. As explained by *Diebler* (*38*), deprotonation of $Hdip^+$ determines the rate of ring closure. In view of the steric constraints, and of the far lower acidity of porphyrins H_2P, it is not surprising that in Eq. (8) k_2 is smaller than in Eq. (10) by a factor of $\sim 10^{-5}$. Hence, if access of at least one pyrrole N to the metal is anticipated, an essential barrier to further substitution at Cu_{aq}^{2+} persists due to the protons on pyrrole nitrogens. The importance of deprotonation as a rate-determining factor was stressed rather early by *Fallab* and *Bernauer* (*39, 40*) who studied the kinetics of metal insertion into phthalocyaninetetrasulphonate. In this context, it is worth commenting upon the pre-equilibrium involving the free base H_2P and the mono anion HP^-:

$$H_2P \; \rightleftharpoons \; HP^- + H^+ . \tag{14}$$

At pH 3, for $H_2P = TMPyP(+4)$, the concentration of HP^- equals $10^{-10} \cdot [H_2P]$. If one makes the assumption

$$\frac{d[CuP]}{dt} = k_2' \cdot [Cu^{2+}] [HP^-] \tag{15}$$

one obtains $k_2' \approx 10^{+10}$ $M^{-1}s^{-1}$, which is comparable to $k_2 \approx 4 \cdot 10^{+9}$ as found for the formation of the ethylenediamine complex $Cu(en)^{2+}$. More experimental evidence will be provided in Section 3, which rules out significant contributions from terms of the type in Eq. (15). Whereas all the factors discussed above are strongly dependent on Λ in $M\Lambda$, the flexibility of H_2P should not be significantly influenced by the particular $M\Lambda$.

2.3.3. Some Remarks on Rate Laws. Experimentally, formation of metalloporphyrins has always been followed spectrophotometrically. Due to high extinction coefficients in the visible, the concentration of porphyrin is usually kept below 10^{-4} M. Thus, it is possible to choose total concentrations of metal species, $[M]_t$, in considerable excess of total porphyrin $[P]_t$. In these cases first-order dependence according to

$$R = \frac{d[MP]}{dt} = k_e \cdot [H_2P] \tag{16}$$

is observed in buffered solutions.

$$k_e = k_2 [M\Lambda] \tag{17}$$

Quite often Eq. (17) has been verified. This linear relation is consistent with the simplified scheme (Eqs. 11) and (12) in Section 2.3.2, provided $K_0 < 10$ and $[M]_t < 10^{-2}$ M. On the other hand, $K_0 > 10$ would cause deviations from the linearly dependent Eq. (17), which then would transform into

$$R = k_z \cdot \frac{K_0 [M\Lambda]}{1 + K_0 [M\Lambda]} \cdot [H_2P]' \tag{18}$$

where

$$\begin{aligned} [H_2P]' &= [H_2P] + [(M\Lambda, H_2P)] \\ &= [H_2P] (1 + [M\Lambda] \cdot K_0) . \end{aligned} \tag{19}$$

As long as $[M\Lambda]$ is well in excess of $[P]_t$, the two components determining the spectral absorption are $[MP]$ and $[HP]'$, the latter comprising the sum (19). Experimentally, a check on Eq. (18) is based on varying $(M\Lambda)$ to a sufficient degree.

135

Hence, the composition of the bulk phase will undergo significant changes unless there is quite a large concentration of some inert electrolyte. This well-known principle, which has been applied to various types of solution studies, will not help if the porphyrin associates selectively with certain components of the solution. In such case, K_0 may not be a constant parameter over the entire range of $[M\Lambda]$ studied. Two specific examples will be considered to illustrate the problems that may arise in evaluating the dependence of k_e on components of the solution.

The insertion of Cu(II) from imidazole (im) complexes into TMPyP($+4$) has been studied in solutions containing the series $Cu(im)_n^{2+}$, $0 < n < 4$. It could be expected (1) that the individual members in the series would react at different rates, and (2) that for constant pH, $\bar{n}$, $[Cu(II)]_t$ the observable parameter k_e would depend on ionic strength. It is seen from Fig. 4 that

(i) for constant $\bar{n}$, pH and $[im] = [imH^+]$, k_e is not proportional to $[Cu(II)]_t$, nor is Eq. (18) valid;

(ii) rates do not change by more than 30% if $\bar{n}$ is increased from 2.3 to 3.9;

(iii) ionic strength effects seem to be screened off.

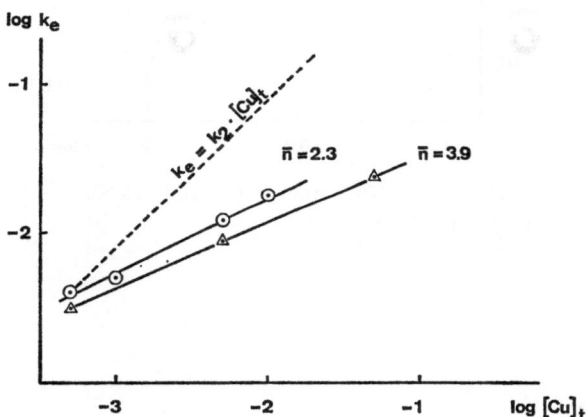

Fig. 4. Insertion of Cu(II) into TMPyP from imidazole complexes $Cu(im)_n^{2+}$ in aqueous solution, pH 7; 25 °C. $k_e(sec^{-1})$, $CuCl_2(M)$.
○ I from 0.003 to 0.03; $[Him^+] = [im] = 1.5 \cdot 10^{-3}$ M.
△ I from 0.1 to 0.25; $[Him^+] = [im] = 0.1$ M

Hence, there is no simple dependence of the type

$$k_e = \sum_n k_n [Cu(im)_n] \tag{20}$$

and analysis of kinetic data is virtually impossible without independent information concerning interactions between the porphyrin and the solute components im, Him^+ and $Cu(im)_n^{2+}$. As for their stability, the members of the series

$Cu(NH_3)_n^{2+}$ differ very little from the corresponding imidazole complexes. A linear dependence $k_e = k_2 \cdot [Cu(II)]_t$ has been observed in one molar ammonia/ammonium buffer in which practically all Cu(II) is tetramminecomplex. The rate constant k_2 is about half the value measured at pH 3 in KNO_3 1 M. Fig. 4 shows the effect of increasing concentrations of NH_3, NH_4Cl and $NaCl$ on the rates relative to a buffer solution containing equal concentrations of NH_3 and NH_4Cl (0.1 M).

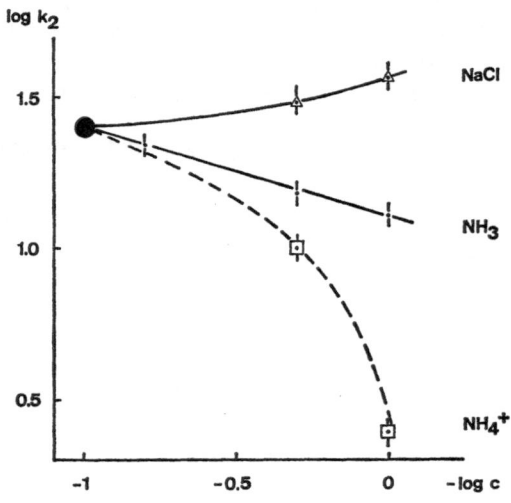

Fig. 5. Insertion of Cu(II) into TMPyP from $Cu(NH_3)_4^{2+}$; 25 °C.
● : $[NH_4^+] = [NH_3] = [Cl^-] = 0.1$ M; $I = 0.1$
△ : $[NH_4^+] = [Na^+] = [Cl^-] = c$; $I = c$
⫯ : $[NH_3] = c$; $I = 0.1$
□ : $[NH_4^+] = c$; $I = c$

It is clear that

(i) increasing ionic strength I from 0.1 to 1.0, together with considerable changes in the medium, $i.e.$ NH_4Cl 0.1 M → NH_4Cl 0.1 M, $NaCl$ 0.9 M, causes only minor changes in k_e;

(ii) passing to $I = 1$ M with NH_4Cl reduces k_e by a factor of about 0.1;

(iii) increasing $[NH_3]$ to 1 M, hence raising pH by one unit, lowers k_e by a factor of 2.

These observations indicate that the variables determining k_e are interdependent, the major reason being a specific ion effect of NH_4^+ which probably undergoes attractive interactions with the central part of TMPyP(+4). Consequently, NH_4^+ and Na^+ are not at all equivalent with respect to short-range interactions with porphyrins. There are other examples indicating competition between (i) ions of the same sign, $i.e.$ $NH_4^+ > Na^+$, $Li^+ > Na^+$, CH_3—C_6H_4—$SO_3^- >$

NO_3^-, Cl^- and (ii) ions of opposite sign, $NH_4^+ > NO_3^-$. Rates may be varied by several orders of magnitude simply by replacing $Na^+ \rightarrow NH_4^+$, $Cl^- \rightarrow Ts^-$ in the reaction discussed above (Table 4).

Table 4. Salt effects in the reaction of $Cu(NH_3)_4^{2+}$ with TMPyP($+4$). Composition of solutions: TMPyP $5 \cdot 10^{-5}$ M, NH_3 0.1 M, ionic strength 1 M (MX). 25 °C. k_2 ($M^{-1}sec^{-1}$). Ts^-: Paratoluenesulfonate

MX (M)	NH_4Cl 0.1 NaCl 0.9	NH_4Cl 1.0	NH_4Ts 1.0
k_2	25	2.4	0.01
k_2 (rel)	1	0.1	0.0004

These examples demonstrate that in certain cases rate laws obtained by data-fitting may be doubtful, and in particular that rate constants emerging from satisfactory fits must be examined with care if they refer to different Λ and the same M in $M\Lambda$. The effects of varying Λ, on the other hand, provide the most valuable information on the mechanism of metalloporphyrin formation.

Temperature dependence has been determined in a variety of cases. However, activation parameters such as A and E_A do not provide as much information about activated states as they do in simpler substitution reactions.

2.4. Metals Selected for Kinetic Studies

From the series of transition-metal ions

$$Mn(II) \quad Fe(II) \quad Co(II) \quad Ni(II) \quad Cu(II) \quad Zn(II) \tag{21}$$
$$Fe(III)$$

or which a wealth of thermodynamic and kinetic data are available (30, 41) Cu(II) and Zn(II) have received most attention. It is only recently that data on Fe(III), Ni(II) and Cu(I) have been obtained (5). Under comparable conditions, the order

$$Cu(II) > Zn(II) > Co(II) > Mn(II) > Ni(II) \tag{22}$$

has been observed for the rates of insertion of hydrated metal ions into TMPyP at pH 4, 50 °C (42). The rates of water exchange follow the same order. Very recently, rates of incorporation from metal halides into TPP were determined in dimethylformamide at 70 °C (43).

Unlike (22), the series

$$Cu(II) > Zn(II) \approx Pd(II) > Co(II) \approx Ni(II) > Mg(II) \tag{23}$$

refers to complexes having widely different Λ, as shown by the significantly different positions of Zn(II) and Pd(II) and also of Co(II) and Ni(II), respectively. The activation enthalpy ΔH^{+} is larger for Cu(II) (16.3 kcal) than for Pd(II) (14.8 kcal). Obviously, variation of Λ may overcome inherent differences in the lability of metal ions. It is relevant to recall here the ranges of second-order rate constants for substitution reactions involving monodentate, or open-chain multidentate ligands (Fig. 6).

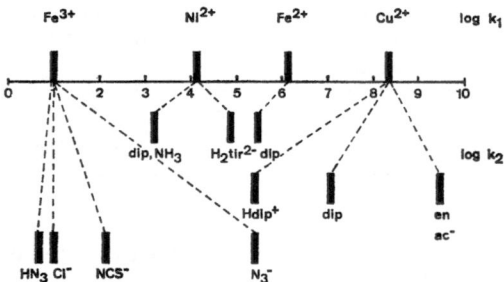

Fig. 6. Rate constants for water exchange of aquo complexes, k_1, and second-order rate constants, k_2, for the formation of complexes ML (*30, 37*)

Hydrated Fe^{3+} and Cu^{2+} differ widely in their rates of water exchange, and there is a spread of rates of substitution for each of these ions over several orders of magnitude. As will be described in Section 3, the large differences in k_1 between hydrated Fe^{3+}, Ni^{2+} and Cu^{2+} do not prevent other complexes of Fe(III) and Ni(II) from reacting with TMPyP only about 10^{-2} times more slowly, or equally fast as Cu^{2+}.

The series (22) represents a rather rare case of a set of data obtainable under really comparable circumstances. Fe^{3+}_{aq} could not be studied at pH 4, but only at pH ≤ 1, where all porphyrins are protonated. As for series (23), quite different species are present in solutions of the corresponding halides in DMF and other organic solvents. As was concluded in Section 2.3.3, comparison of rate data on a set of MΛ differing in either M and/or Λ necessitates rigorous analysis of individual systems.

3. Kinetic Studies in Aqueous Solution

3.1. General and Specific Salt Effects

Inert salts have been introduced either to maintain constant ionic strength, or for studying salt effects. The major part of the relevant data refers to the insertion of hydrated Cu^{2+} and Zn^{2+} into TMPyP($+4$). For both metal ions, a linear dependence on nitrate concentration

$$R = k_3 \, [M^{2+}] \, [TMPyPH_2] \, [NO_3] \qquad (24)$$

has been observed in the interval $0.2 < MNO_3 < 2$ M, *i.e.* at rather high ionic strength I *(11, 44)*. The rate law (24) is restricted to $I > 0.1$, as shown by data *(5)* on Cu^{2+} at rather low I. According to *Hambright* *(44)*, $k_3[NO_3]$ equals 0.23 $M^{-1}s^{-1}$ (26 °C) in $NaNO_3$ 0.1 M, but the second-order rate constants for CuX_2 obtained in the interval $0.002 < I < 0.1$ (NaX, $X^- = NO_3^-$, Cl^-, Ts^-) extrapolate $(I \to 0)$ to a value of 0.25 (± 0.05). Salt effects of KNO_3, $NaNO_3$ and $NaCl$ produce increased rates above $I = 0.1$ but the effects are clearly larger for Cl^- than for NO_3, *i.e.* $k_3(NaCl) \approx 3 \cdot k_3(NaNO_3)$. The standard inert ion ClO_4^- unfortunately precipitates $TMPyP(ClO_4)_4$. Paratoluenesulphonate (Ts^-), like Cl^-, NO_3^- and ClO_4^-, does not form complexes with Cu^{2+}. However, decreased rates are caused by $NaTs$ at concentrations as low as 10^{-3} M. As shown in Fig. 7, the concentration of Ts^- determines the decrease in rate, irrespective of whether the cation is Na^+ or Ca^{2+} *(6, 7)*.

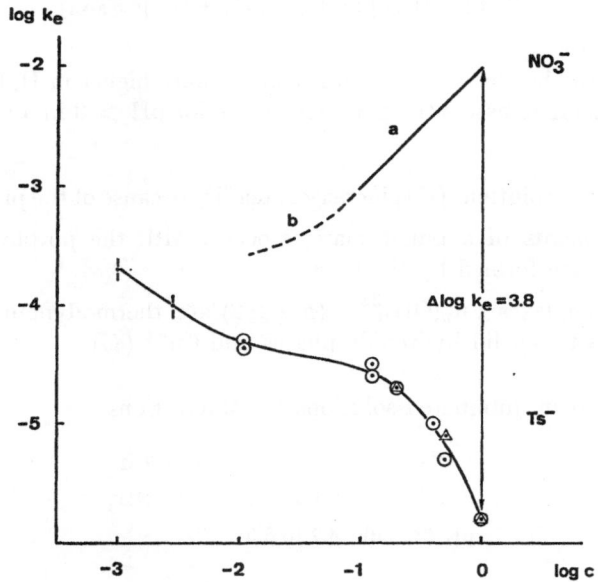

Fig. 7. Salt effects in the reaction of Cu_{aq}^{2+} with TMPyP($+4$), pH 3; 25 °C. $[Cu^{2+}] = 5 \cdot 10^{-3}$ M; $k_e(sec^{1-})$. $Ts^- = CH_3-C_6H_4-SO_3^-$
a : KNO_3, Ref. *(11)*; *b*: KNO_3, Ref. *(5)*
⌶ : $Cu(NO_3)_2$, HNO_3, $NaTs$
○ : $CaTs_2$
△ : $NaTs$

H-NMR studies reveal unambiguously that TMPyP($+4$) associates quite effectively with Ts^- *(9)*. Moreover, inhibiting influences of tosylate have been confirmed for Λ containing ligands such as NH_3, N_3^- and HCO_2^- *(6)*. It must be concluded that the long-range effects of charges are less important than the solvation domain close to the porphyrin. It has not been established whether micelles involving more than one porphyrin molecule are formed. Similar association phenomena have been found for the dianion of tiron (4,5-dihydroxyben-zene-1,3-disulphonic acid) and for sulphosalicylate (9). According to *Hambright* (11), any general salt effect of $NaNO_3$ is screened off by PIPES buffer in the reaction of Zn_{aq}^{2+} with $TPPS_4(-4)$.

Summarizing available evidence at present, specific ion effects are hardly separable from general salt effects in alkali halides or nitrates, but may easily become dominant due to short-range interactions in the presence of larger ions.

3.2. Complexes of Cu(II) and Cu(I)

Lowering the pH at constant ionic strength produces decreasing rates of insertion of Cu_{aq}^{2+} due to the lower fraction of TMPyP present in the free-base form H_2P in accordance with

$$\sum_{i=2}^{4} [H_iP] = [H_2P] \{1 + [H^+] K_3 + [H^+]^2 K_3K_4\}. \tag{25}$$

Obviously the barrier to insertion is considerably higher in H_3P^+ and H_4P^{2+}. It is quite difficult to assess the pH dependence for $pH > 3$ in a reliable way for three reasons:

(i) in unbuffered solutions $[H^+]$ increases steadily because of the protons released;

(ii) the components of a buffer may associate with the porphyrin, or Cu(II) complexes are formed by the base;

(iii) hydroxocomplexes $Cu_p(OH)_q^{2p-q}$ ($p, q \geq 2$) are thermodynamically unstable with respect to solid hydroxide phases and Cu^{2+} (45).

From studies in unbuffered solutions of compositions

$$\begin{array}{ll} CuX_2 & 5 \cdot 10^{-4} \text{ to } 5 \cdot 10^{-3} \text{ M} \\ KX & 0.1 \text{ M} \quad X^- = Cl^-, NO_3^- \\ pH \text{ (initial)} & 4.7 \text{ to } 5.0 \end{array} \tag{26}$$

unambiguous proof has been obtained that some hydroxocomplex of Cu(II) must react several orders of magnitude faster than Cu_{aq}^{2+} (46). Solutions of type (26) contain hydroxocomplexes which change over a period of days into less reactive species.
The function

$$\zeta(t) = -\frac{1}{[H_2P]} \cdot \frac{d[H_2P]}{dt}. \tag{27}$$

has been calculated from experimental data as a continuous function of time. Considering

$$[Cu^{2+}] \gg ([H_2P] + [CuP]) \tag{28}$$

$$[Cu^{2+}] \gg \sum_{p, q} p [Cu_p(OH)_q] \tag{29}$$

141

it is reasonable to use the normalized function

$$\frac{\zeta(t)}{[Cu^{2+}]} = \chi(t) \tag{30}$$

for comparing with values obtained at pH 3 where χ is the second-order rate constant k_2. As can be seen from Fig. 8, $\chi(t)$ is considerably larger than k_2 and decreases as $[H^+]$ increases to its final value

$$[H^+]_\infty = [H^+]_0 + 2\,[CuP]_\infty . \tag{31)[4]}$$

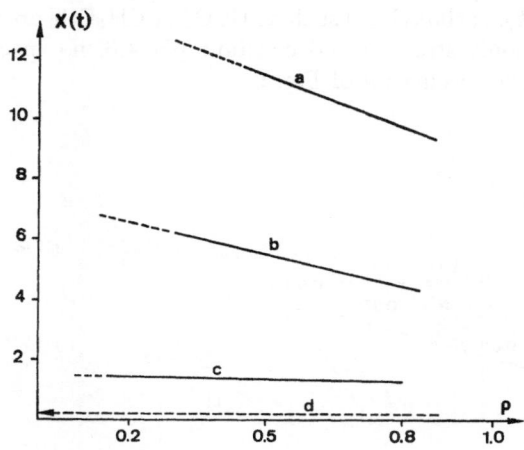

Fig. 8. Insertion of Cu(II) into TMPyP from unbuffered solutions (26); KNO_3 0.1 M; 25 °C. The function $\chi(t)$, Eq. (30), is plotted vs the fraction ϱ of $Cu^{II}P$ formed. Initial concentrations:
a: $Cu(NO_3)_2$ $5 \cdot 10^{-4}$ M; TMPyP $2.1 \cdot 10^{-5}$ M
b: $Cu(NO_3)_2$ $1 \cdot 10^{-3}$ M; TMPyP $2.05 \cdot 10^{-5}$ M
c: $Cu(NO_3)_2$ $2 \cdot 10^{-3}$ M; TMPyP $2.36 \cdot 10^{-5}$ M.
   Age of stock solution 2d ($Cu(NO_3)_2$ $4 \cdot 10^{-3}$ M)
d: second-order rate constant k_2 for pH $= 3$

It is well known that on the time scale of the insertion reaction the protonation of $Cu(OH)^+$ and $Cu_2(OH)_2^{2+}$ is a rapid reaction. If pre-equilibria involving these hydroxocomplexes were established continuously and the subsequent formation of higher polynuclear species proceeded much more slowly, one could apply the rate law:

$$\zeta(t) = k_2\,[Cu^{2+}] + k_2'\,[Cu(OH)^+] + k_2''\,[Cu_2(OH)_2^{2+}]$$
$$K_1^* = \frac{[Cu^{2+}]}{[Cu(OH)^+]\,[H^+]} \; ; \quad \beta_{22}^* = \frac{[Cu^{2+}]^2}{[Cu_2(OH)_2^{2+}]\,[H^+]^2} \tag{32}$$

---

[4]) Equation (31) refers to analytical concentrations of H^+.

Inserting values suggested by literature data for K_1^* (47) and β_{22}^* (48), curve-fitting procedures have been applied to test the validity of Eq. (32). Although each individual run involving k_2 obtained at pH 3 could be fitted with Eq. (32), the pairs of parameters k_2' and k_2'' depend on $[Cu]_t$ and on the age of the solution. Hence, the conditions for applying Eq. (32) are never fulfilled in a strict sense. Independent evidence is given below to support the conclusion that $Cu(OH)^+$ is the kinetically relevant hydroxocomplex.

If we accept $7 < \log K_1^* < 8$, data such as those shown in Fig. 8 are compatible with

$$Cu(OH)^+ : 10^{+3} < k_2' < 10^{+4} \text{ M}^{-1}\text{s}^{-1}. \tag{33}$$

Whereas catalysis by OH^- coordinated to $Cu(II)$ produces effects of the order of at least 10^{+4} in k_2, carboxylates such as HCO_2^- or $CH_3CO_2^-$ are much less effective. At rather high ionic strength and constant pH 4.6, no spectacular catalysis is observed as shown in curve a of Fig. 9.

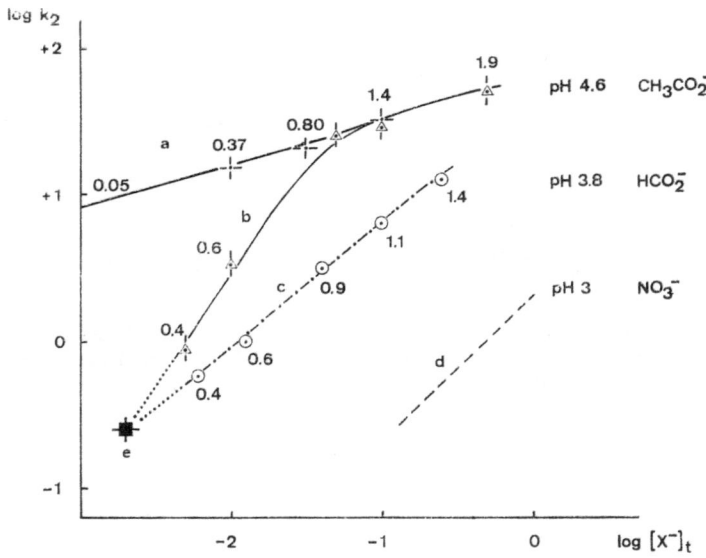

Fig. 9. Catalysis of the insertion of Cu(II) into TMPyP by acetate and formiate. $k_e(\text{sec}^{-1})$; 25 °C. (5, 6, 7). Average number $\bar{n}$ of carboxylates per Cu(II) indicated.
a: $I = 1.0$ (KNO₃); $[Cu]_t$ $1 \cdot 10^{-4}$ M
b: I from 0.006 to 0.50 (NaCH₃CO₂); $[Cu]_t$ $5 \cdot 10^{-4}$ M. $k_2 = k_e/[Cu]_t$
c: I from 0.007 to 0.26 (NaHCO₂); $[Cu]_t$ $5 \cdot 10^{-4}$ M. $k_2 = k_e/[Cu]_t$
d: KNO₃, HNO₃, from Ref. (11)
e: k_2 ($I \rightarrow 0$); pH 3

These data are reproducible by a rate law

$$R = k_2^0 [Cu^{2+}] + k_2' [Cu(ac)^+] + k_2'' [Cu(ac)_2] \tag{34}$$

143

where $k_2^0 \approx k_2' < k_2''$. Actually, k_2^0 accounts for the small fraction of Cu_{aq}^{2+} which is deprotonated at pH 4.6. Curve b joins curve a at higher concentrations of acetate, thus showing that k_2'' is independent of I opposite to k_2^0 and k_2'. In the series b and c of Fig. 9, the ionic strength is determined by the concentration of $NaRCO_2$ ($R = CH_3$, H). Comparison with data for pH 3 in curve d unambiguously demonstrates catalysis by RCO_2^-. Rate constants for $Cu(RCO_2)_2$ have been derived from data for higher concentrations of RCO_2^-:

$$Cu(CH_3CO_2)_2 : k_2'' \approx 50 \ M^{-1}s^{-1} \ (25 \ °C)$$
$$Cu(HCO_2)_2 \quad : k_2'' \approx 30 \ M^{-1}s^{-1} \ (25 \ °C) \tag{35}$$

In Fig. 9, curves b and c extrapolate to values around point e for $[RCO_2^-] \approx 2 \cdot 10^{-3}$ M, where $[Cu(RCO_2)_2]$ is negligible compared to $[Cu^{2+}]$ and $[Cu(RCO_2)^+]$. This is remarkable, because the pH values for b, c and e are 4.6, 3.8 and 3, respectively, suggesting that in solutions containing $[RCO_2^-] = [RCO_2H]$ there is no dependence of k_2^0 on pH. When all other variables are kept constant, increasing concentration of CH_3CO_2H lowers the rates of insertion of Cu(II) or Ni(II) (7), suggesting selective association of TMPyP with CH_3CO_2H. Clearly, rate laws such as (34) cannot be valid in a strict sense because the variables are not really separable.

Azide is more effective than acetate, as shown clearly by data obtained in 0.2 M Na_2Ts (Fig. 10). From the rate dependence on $[Cu(N_3)^+]$ calculated from equilibrium data (41) the relevant rate constant can be estimated to be $k_2 \approx 20$ $M^{-1}s^{-1}$, which is considerably higher than $k_e/[Cu]_{tot} = 0.04$ for the corresponding system without azide. Correcting for the inhibiting effect of tosylate, it is deduced

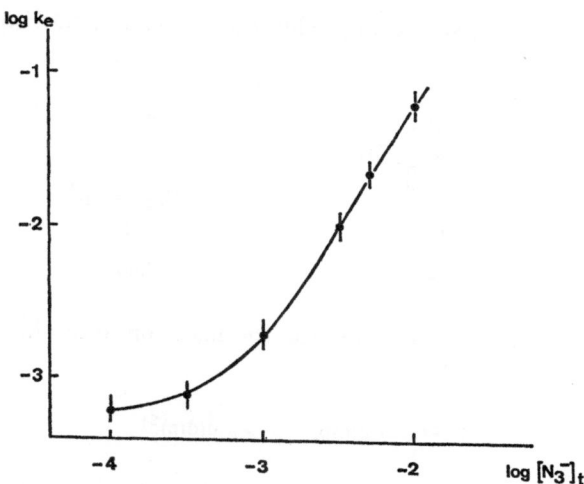

Fig. 10. Catalysis of the insertion of Cu(II) into TMPyP by azide. Total concentrations: Cu(II) $5 \cdot 10^{-3}$ M; CH_3CO_2H, $CH_3CO_2^-$ $5 \cdot 10^{-3}$ M; $I = 0.2$ (NaTs); 25 °C; k_e(sec^{-1})

that under comparable conditions the relative rates of insertion of Cu^{2+}, $Cu(N_3)^+$ and $CuOH^+$ cover a range of at least three orders of magnitude:

$$Cu^{2+} \quad Cu(N_3)^+ \quad Cu(OH)^+ \qquad (36)$$
$$k_2 \text{ (rel.)} \quad 1 \qquad 5 \cdot 10^{+2} \qquad \geq 2.5 \cdot 10^{+3} .$$

In this context, it is quite important that $Cu(NH_3)^{2+}$ reacts only slightly faster than Cu^{2+} under comparable conditions. The kinetic data plotted in Fig. 11 demonstrate that the species $Cu(NH_3)_n^{2+}$, $n \geq 2$, reacts more slowly than $Cu(NH_3)^{2+}$.

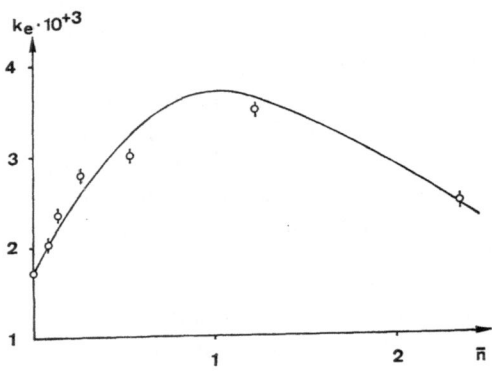

Fig. 11. Insertion of $Cu(NH_3)_n^{2+}$ into TMPyP. Total concentrations: Cu(II) $1 \cdot 10^{-3}$ M; NH_4NO_3 1 M; 25 °C. pH: 2 (HNO₃, $\bar{n} = 0$); 4 to 6.4 ($\bar{n}$ 0.09 to 2.34).

As for bidentate ligands, spectacular effects are produced by tironate (IX) and 8-oxyquinolinate (X).

IX

H_2tir^{2-}

X

hoq

On the other hand, dipyridyl retards the insertion according to rate coefficients decreasing in the order

$$Cu^{2+} > Cu(dip)^{2+} >> Cu(dip)_2^{2+} . \qquad (37)$$

The insertion of Cu(II) from $Cu(tir)^{2-}$ into TMPyP($+4$) in the interval $3.4 < pH < 3.9$ had to be followed by the stopped-flow method (7). The data in Fig. 12 show that the reaction is slightly retarded by an excess of H_2tir^{2-}, which forms complexes $[\text{TMPyP} \cdot (\text{H}_2\text{tir})_x]^{4-2x}$. This has also been observed in the insertion of Fe(III) from Fe(tir)$^-$ and Fe(tir)$_2^{5-}$ (Section 3.5), the effects being compatible with $x = 1$, $K \sim 1.2$ (± 0.3) $\cdot$ 10^{+2} in all cases.

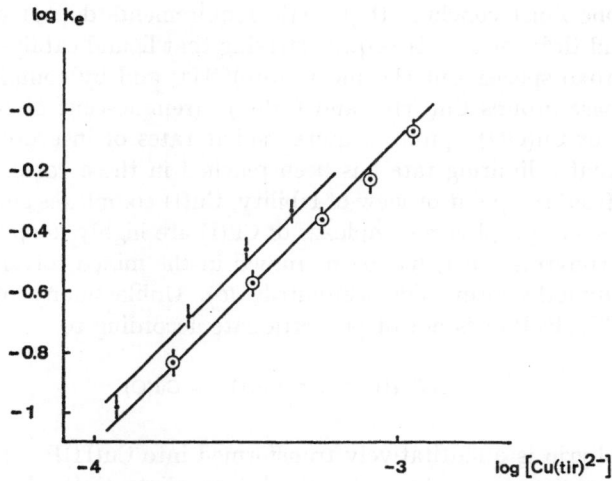

Fig. 12. The reaction of Cu(tir)$^{2-}$ with TMPyP in aqueous solution, $I = 0.1$ (KNO$_3$); 25 °C. Observed rate coefficient k_e vs. concentration of Cu(tir)$^{2-}$ calculated from equilibria data (41)

Table 5. The influence of bidentate ligands on the insertion of Cu(II) into TMPyP. Second-order rate constants k_2 (M^{-1}sec^{-1}) refer to complexes in the first column. 25 °C. H$_2$tir^{2-} (IX); hoq (X); dip: 2,2'-dipyridyl

Complex	k_2	Medium		Ref.
Cu(tir)$^{2-}$	$1.6 \cdot 10^{+3}$	KNO$_3$	0.1 M; pH 3.3—3.9	(5, 7)
Cu(oq)$^+$	$0.7 \cdot 10^{+3}$	Na$_2$SO$_4$	0.1 M; pH4	(42)
Cu$_{aq}^{2+}$	0.20	KNO$_3$	0.1 M; pH 3	(7, 11)
Cu(dip)$^{2+}$	0.11	KNO$_3$	0.1 M; pH 3	(7)
Cu(dip)(OH)$^+$	$\sim 2 \cdot 10^{+3}$	KNO$_3$	0.1 M; pH 9.2—10.2	(7)
Cu(dip)(OH)$_2$	$\sim 4 \cdot 10^{+3}$			

Ligand catalysis by oxyquinolate (42) is also very effective (Table 5). The striking difference between the strongly basic bidentate ligands containing one (oq$^-$) or two (tir^{4-}) oxygen atoms on one hand and the nitrogen ligand dip on the other hand reveals the importance of proton transfer from H$_2$P to a coordinated ligand. This is also supported by the observation that the mononuclear Cu(dip)(OH)$^+$ reacts much faster than the binuclear Cu$_2$dip$_2$(OH)$_2^{2+}$, whereas Cudip(OH)$_2$ is equivalent to Cudip(OH)$^+$ within a factor of two. Kinetic data obtained by

146

stopped-flow procedures have been analyzed on the basis of Martell's equilibrium data (41) by applying the rate law

$$R = k_2 \left[\text{Cu(dip)OH}\right] + k_2' \left[\text{Cu(dip)(OH)}_2\right] + k_2'' \left[\text{Cu}_2\text{dip}_2\text{(OH)}_2\right]. \tag{38}$$

Hence, one must conclude that steric requirements do not discriminate between dip and tir^{4-}, or oq$^-$. It is quite striking that ligand catalysis by the mononuclear hydroxo species CuOH$^+$ and Cu(dip)OH$^+$, and by complexes containing the strong base groups Cu(tir)$^{2-}$ and Cu(oq)$^+$ (reminescent of the hypothetical cis-Cu(OH)$_2$ or Cu(OH)$^+$ produce quite similar rates of insertion. It is justified to suppose that a limiting rate has been reached in these cases.

Clearly, from the point of view of lability, Cu(I) complexes should react fairly fast. As expected, porphyrin complexes of Cu(I) are highly unstable with respect to Cu(II) porphyrins. Cu(I) has been studied in the mixed solvents water-acetonitrile and dimethylformamide-acetonitrile (6). Unlike many other complexes of Cu(I), Cu^{I}TMPyP does not disproportionate according to

$$2\,\text{Cu(I)} \;\rightleftharpoons\; \text{Cu(II)} + \text{Cu(O)}. \tag{39}$$

The porphyrin is quantitatively transformed into Cu(II)P, which means that solvent components are reduced by the intermediate Cu^{I}P. Hence, observable rates of formation of Cu^{II}P refer to concerted insertion and electron-transfer steps. The dependence on total Cu(I) in CH$_3$CN 5 M corresponds to a case of 1:1 association with stability constant $K \approx 10^{+3}$, as indicated in Fig. 13.

When the rate law (18) is applied, the product $(k_2 \cdot K_0) \approx 3 \cdot 10^{+2}$ M^{-1}s^{-1} is obtained, showing that in water-acetonitrile (5 M) insertion of Cu(I) proceeds more slowly than insertion of Cu(II) from Cu(tir)$^{2-}$ described above. However,

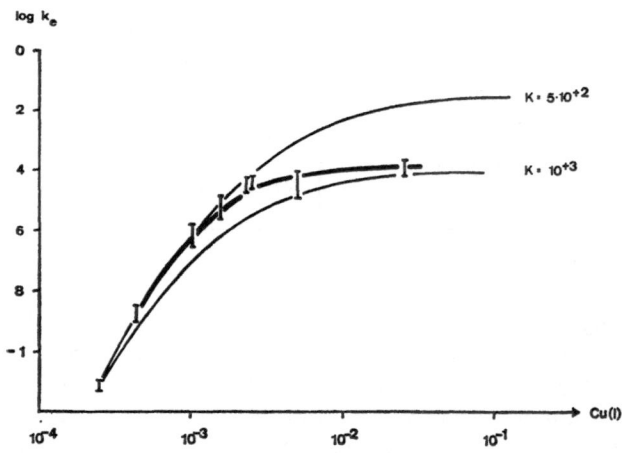

Fig. 13. The reaction of Cu(NCCH$_3$)$_n^+$ with TMPyP in water-acetonitrile (5 M); 25 °C. Calculated curves represent the law (18) with association constants indicated

the rates depend strongly on the concentration of CH_3CN. At the lower limit, about 1 M CH_3CN, Cu(I) is inserted slightly faster than $Cu(tir)^{2-}$. Increasing concentration of CH_3CN favours the higher members in the series $Cu(NCCH_3)_n^+$, $n = 2, 3, 4$ (49). As indicated in Fig. 14, tetrahedral $Cu(NCCH_3)_4^+$ reacts at least

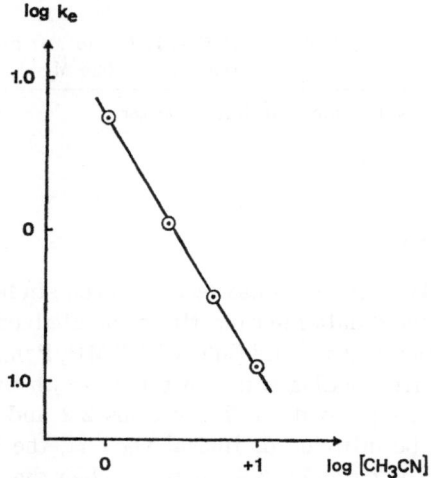

Fig. 14. The dependence of k_e on the concentration of CH_3CN. $[Cu(I)] = 5 \cdot 10^{-3}$ M

two orders of magnitude more slowly than $Cu(NCCH_3)_2^+$. The imidazole complex $Cu(im)_2^+$ is far more stable than $Cu(NCCH_3)_2^+$, overall stabilities amounting to $\log \beta_2$ 10.8 and 4.3, respectively (49). From the linear dependence of k_e on $Cu(im)_2^+$, $k_2 = 5 \cdot 10^{+2}$ $M^{-1}s^{-1}$ has been determined, providing another example of complexes for which $k_2 \sim 10^{+3}$ $M^{-1}s^{-1}$. Table 6 summarizes data on Cu(I) and Cu(II).

Table 6. The influence of monodentate ligands on the insertion of Cu(II) and Cu(I) into TMPyP. k_2 $(M^{-1}sec^{-1})$; 25 °C; aqueous solutions

Reactants	k_2	Medium	Ref.[c]
Cu(II)-TMPyP			
Cu_{aq}^{2+}	2.3	$NaNO_3$ 1 M; pH 3	(44)
	7.0	Na_2SO_4 0.1 M; pH 3–4; 55 °C	(42)
$Cu(NH_3)^{2+}$	5	NH_4NO_3 1 M; pH 4–6	(5, 7)
$Cu(NH_3)_{2,3,4}^{2+}$	2		
$Cu(N_3)^+$	20	NaTs 0.2 M; acetate buffer	(6)
Cu^{2+}	0.04	NaTs 0.2 M; acetate buffer	(5, 7)
$Cu(OH)^+$ est.	$\geqslant 10^{+3}$	$Na(Cl,NO_3)$ 0.1 M	(46)
$Cu(CH_3CO_2)_2$	50	I var., $M(Cl^-, NO_3^-)$	(5)
Cu(II)-Deutero[a]			
$Cu(CH_3CO_2)_n^{2-n}$	3.2	I 0.5; CH_3CO_2 (H,Na) 0.1 M	(16)

Table 6. continued

Reactants	k_2	Medium	Ref.[c]
Cu(II)-Haemato[b]			
Cu(II)	0.02	NaOH 0.5 M	(16)
Cu(I)-TMPyP			
Cu(im)$_2^+$	$5 \cdot 10^{+2}$	H$_2$O—CH$_3$CN (5 M); pH $\sim$7 Him$^+$ / im 0.025 M	(6)

[a] Deuteroporphyrin-2,4-disulphonic acid dimethylester.
[b] Haematoporphyrin IX.
[c] For further data up to 1973, see Ref. (43).

3.3. Complexes of Zn(II)

Variation of Λ in Zn(II) complexes has recently been studied by *Hambright* (11), whereas previous scattered data refer mostly to solvated ions. The same rate laws are found for the reaction of Zn^{2+} and Cu^{2+} with TMPyP in alkali nitrate solutions at pH <4, but the corresponding rate constants are about 50 times smaller for Zn^{2+}. The complications pointed out in Sections 2.2 and 2.3.3 arise at pH >4 when buffers have to be introduced. In NaNO$_3$ 1 M, the 2,6-lutidine buffer enhances the rates, and the parameters k_e increase when the pH is raised. Catalysis by pyridine has been analyzed (11) in terms of the rate law

$$k_e = \sum_{n=0}^{3} k_n \left[\text{Zn(py)}_n^{2+}\right] [\text{NO}_3] \tag{40}$$

The data show clearly that Zn(py)$^{2+}$ and possibly Zn(py)$_2^{2+}$ react faster than Zn^{2+} by factors of 10 and 40, respectively. Acetate also produces acceleration, as expected from data on Cu(II). Rate expressions of type (40) do not seem to fit data on Zn(NH$_3$)$_n^{2+}$. In particular, no appreciable dependence on nitrate was observed (11). This agrees with the results on Cu(NH$_3$)$_4^{2+}$ discussed in Section 2.3.3. It is less surprising that no effect was observed in the reaction of Zn(NH$_3$)$_n^{2+}$ with TPPS$_4$(-4) when the concentration of NaNO$_3$ was raised from 0.3 to 1.2 M. Second-order rate constants attributed to Zn(NH$_3$)$_3^{2+}$ and Zn(NH$_3$)$_4^{2+}$ are 200 and 10 times larger (8.7 $<$ pH $<$ 9.9) than for Zn^{2+} at pH 6.5. Effects of NH$_4^+$, however, have not been considered in this work (11). If it is true that the rate does not depend on NH$_4^+$, Zn(II) shows quite different behaviour from Cu(II) in amine complexes (see Section 3.2). Even larger effects have been reported in the reaction of Zn(NH$_3$)$_3^{2+}$ with uroporphyrin (11). It is very interesting that Zn(OH)$_3^-$ and Zn(OH)$_4^{2-}$ seem to react with the same porphyrin more slowly than Zn(NH$_3$)$_3^{2+}$ by at least a factor of 10^{-3} (11). However, some reservation may be justified in view of the difficulties of obtaining reliable data on hydroxocomplexes of metal ions quite generally. Polynuclear complexes of Zn(II) would certainly react much more slowly than mononuclear species, as suggested by the behaviour of Cu(II) described earlier.

At pH $\sim$ 7, rates of insertion into TMPyP($+$4) and into the stronger base TPPS$_4$(-4) are about the same in 1 M NaNO$_3$, and 1 M NaClO$_4$, respectively

(Table 7). Hence, no evidence for pronounced effects of differing base strength of pyrrole nitrogens emerges from these data.

Table 7. The insertion of Zn(II) into water-soluble porphyrins. Data from Ref. (*11*)

Reactants	k_2	Medium	Temp.
TMPyP(+4)			
Zn²⁺	0.049	NaNO₃ 1 M; pH 3	22 °C
	0.50	NaNO₃ 1 M; pH 6.7	22 °C
		Lutidine buffer 0.05 M	
Zn(py)²⁺	0.7	NaNO₃ 1 M; pH 5.3	27 °C
Zn(py)₂²⁺	1.8		
Zn(CH₃CO₂)⁺	0.5	NaNO₃ 1 M; pH 3.7	22 °C
Zn(CH₃CO₂)₂	2.9		
TPPS₄(−4)			
Zn²⁺	0.76	NaNO₃ 0.02 to 0.6 M	25 °C
		pH 6.3 to 7.1	
Zn(OH)⁺	41	PIPES ~10⁻² M	
Zn(NH₃)₃²⁺	200	NaNO₃ 0.3 to 1.2 M	25 °C
		pH 8.7 to 9.8	

3.4. Complexes of Ni(II)

As was shown by *Phillips* (*42*), Ni²⁺ reacts with TMPyP much more slowly than Cu²⁺ by a factor of at least $3 \cdot 10^{-4}$ at 55 °C. Data from *Hambright* (*11*) indicate about $0.5 \cdot 10^{-4}$ for the same factor at room temperature. Therefore Ni(II) provides a good case for a test of the principles governing catalysis of incorporation. We are not discussing here in detail how $CH_3CO_2^-$, NH_3 and $NH_2CH_2CO_2^-$ raise coefficients $k_e/[Ni(II)]_t$ up to some 10^{-3} M⁻¹s⁻¹ in corresponding buffers. However, two observations are worth mentioning. First, Fig. 15 demonstrates that, if all other components remain at the same concentration, increasing the concentration of CH_3CO_2H lowers the rates. Secondly, varying the average number $\bar{n}$ of NH_3 per Ni(II) from 2.8 to 4 produces slightly lower rate parameters $k_e/[Ni(II)]$, which indicates that the higher complexes $Ni(NH_3)_n^{2+}$ ($n = 4, 5, 6$) are not reacting faster than the lower ones. However, if $\bar{n} = 5.6$ is reached in strong ammonia solution (NH_3 0.7 M; pH ~11.6), the coefficient $k_e/[Ni(II)]$ increases by a factor of about 500 from $7 \cdot 10^{-4}$ (average for $\bar{n} = 2.8$; 4) to 0.4 M⁻¹s⁻¹. Hence, under suitable conditions, Ni(II) can be inserted as rapidly as Cu_{aq}^{2+}. Of course, it is hardly possible to discriminate between the two possibilities: (i) that a small percentage of HP⁻ reacts very fast with $Ni(NH_3)_n^{2+}$, or (ii) that hitherto undetected $Ni(NH_3)_n(OH)^+$ is superior to ammine complexes in the same way as Cu(dip)OH⁺ with respect to Cu(dip)²⁺. Rather well-defined conditions may be realized when

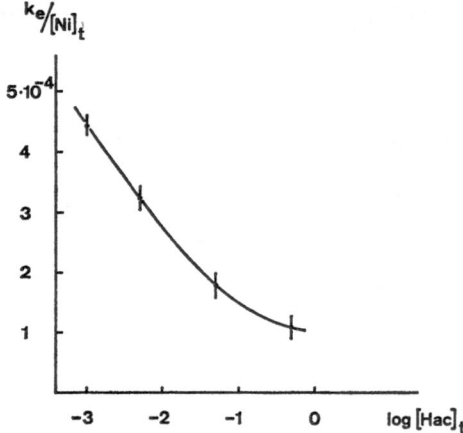

Fig. 15. Formation of Ni(TMPyP)$^{4+}$ in solutions with constant total concentrations Ni(NO$_3$)$_2$ 0.05 M; KNO$_3$ 0.5 M; K(CH$_3$CO$_2$) 0.5 M. Variable: CH$_3$CO$_2$H ($=$ Hac) added. From equilibrium data: Ni(ac)$_n^{2-n}$, $n = 0, 1, 2$; $\bar{n} = 0.93$.

chelating ligands such as glycylglycine (HGG) and glycylsarcosine (HGS) are introduced. Stability constants have been redetermined (*50*) in the system Ni(II)/HGG. The complexes Ni(GGH$_{-1}$) and Ni(GGH$_{-2}$) have been detected by *Martin, Chamberlin* and *Edsall* (*51*) whereas several other groups have determined stabilities of the normal series Ni(GG)$_n^{2-n}$, $n = 1,2,3$ (*52*). Table 8 summarizes the constants used to calculate the composition of reaction mixtures. A typical distribution diagram for a set of solutions studied is shown in Fig. 16. Clearly, the concentrations of Ni(GGH$_{-2}$)$^-$ and Ni(GGH$_{-1}$) change quite considerably from pH 7 to 9.5. In the latter complex, deprotonated glycylglycinate coordinates by N(RNH$_2$), N$^-$(peptide) and O(RCO$_2^-$) to Ni(II). Hence, unless Ni(GGH$_{-2}$) is an artefact of computer fitting procedures, it ought to be a mononuclear hydroxo-complex Ni(GGH$_{-1}$)OH$^-$. It is not present in significant fractions around pH $\sim$7 where k_e/[Ni(II)] approaches similar values (about 10^{-3} M^{-1}s^{-1}), as obtained for

Table 8. Stability constants of complexes in the system Ni(II)-Glycylglycine. I $=0.1$ (KNO$_3$); 25 °C. Ref. (*50*). GG$^-$ = H$_2$NCH$_2$CONHCH$_2$CO$_2^-$

		log K	log β
H$^+$ + GG$^-$	$\rightleftharpoons$ HGG	8.15	
Ni^{2+} + GG$^-$	$\rightleftharpoons$ Ni(GG)$^+$	4.14	
Ni(GG)$^+$ + GG$^-$	$\rightleftharpoons$ Ni(GG)$_2$	3.23	β_2 7.37
Ni(GG)$_2$ + GG$^-$	$\rightleftharpoons$ Ni(GG)$_3^-$	2.54	β_3 9.91
H$^+$ + Ni(GGH$_{-1}$)	$\rightleftharpoons$ Ni(GG)$^+$	8.91	17.64
H$^+$ + Ni(GGH$_{-2}$)$^-$	$\rightleftharpoons$ Ni(GGH$_{-1}$)	8.73	

ammonia, acetato and glycinato complexes. However, from an extended series of kinetic runs, a linear correlation between the experimental parameter k_e and [Ni(GGH$_{-2}$)] emerges as shown in Fig. 17.

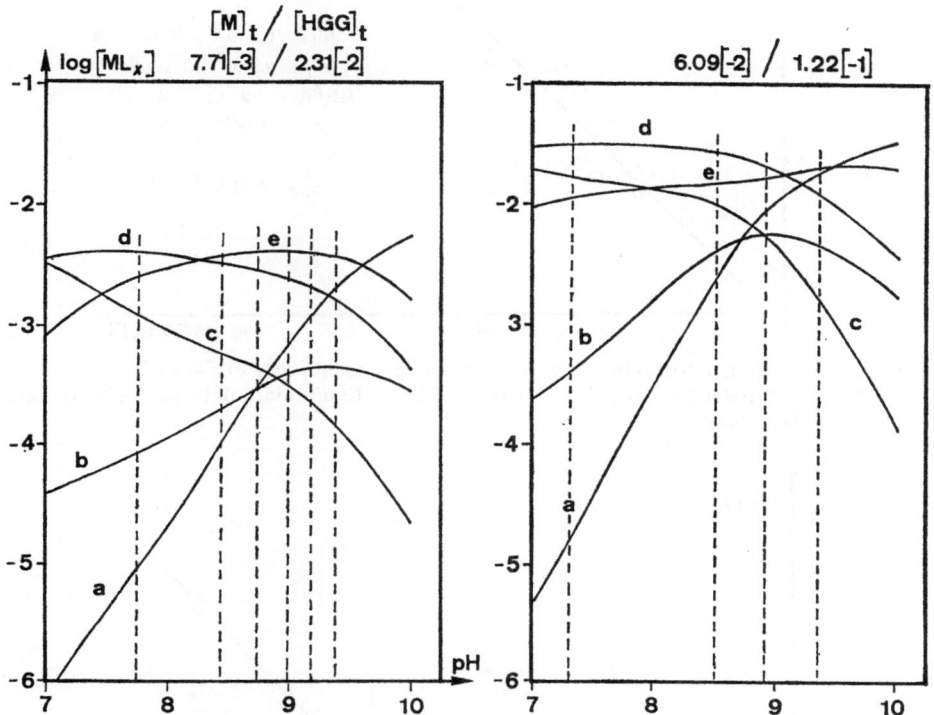

Fig. 16. Distribution diagram of Ni(II) complexes.
HGG = Glycylglycine;
a = Ni(GGH$_{-1}$) (OH$^-$); b = Ni(GGH$_{-1}$)
c = Ni(GG)$^+$; d = Ni(GG)$_2$; e = Ni(GG)$_3$.
Solutions marked with dashed lines were used in kinetic studies (Fig. 17)

The system Ni(II)/glycylsarosine (HGS) is slightly less complicated since no analogue to Ni(GGH$_{-2}$)$^-$ exists. Martin's stability data (51) have been used for this system. The pK 10.9 associated with the pair Ni(GS)$^+$/Ni(GS)H$_{-1}$ is significantly higher than the 8.7 determined for Ni(GGH$_{-1}$)/Ni(GGH$_{-2}$)$^-$. Since GS$^-$ contains a methylated peptide group, it is postulated that Ni(GS)H$_{-1}$ = Ni(GS)OH. From data obtained in the interval $6.9 < \text{pH} < 9.2$, the linear relation in Fig. 18 involves practically the same coefficient $k_2 = 5.7$ (± 1.5) M^{-1}s^{-1} as is derived from the data in Fig. 17. Hence, metal insertion into a porphyrin may provide a method of detecting mononuclear hydroxocomplexes in certain cases. Alternatively, the ligands GGH$_{-1}$ and GS$^-$ are important mainly because they stabilize mononuclear hydroxo complexes with respect to polynuclear species.

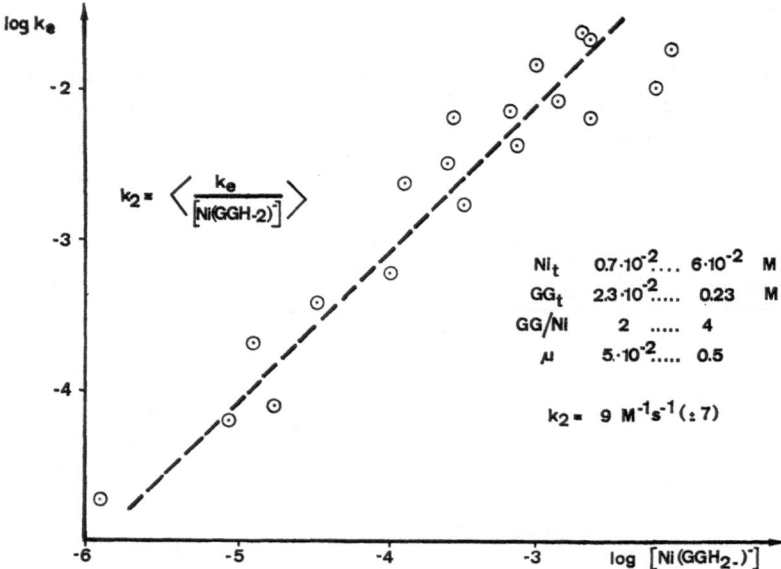

Fig. 17. The insertion of Ni(II) from glycylglycine complexes into TMPyP: k_e vs. $[\mathrm{Ni(GGH_{-2})^-}]$; $\mathrm{Ni(GGH_{-2})^-} = \mathrm{Ni(GGH_{-1})\,(OH)}$. $k_2 = 9 \pm (7)$ $\mathrm{M^{-1}sec^{-1}}$. Equilibrium data for ionic strength $I = 0.1$ have been used

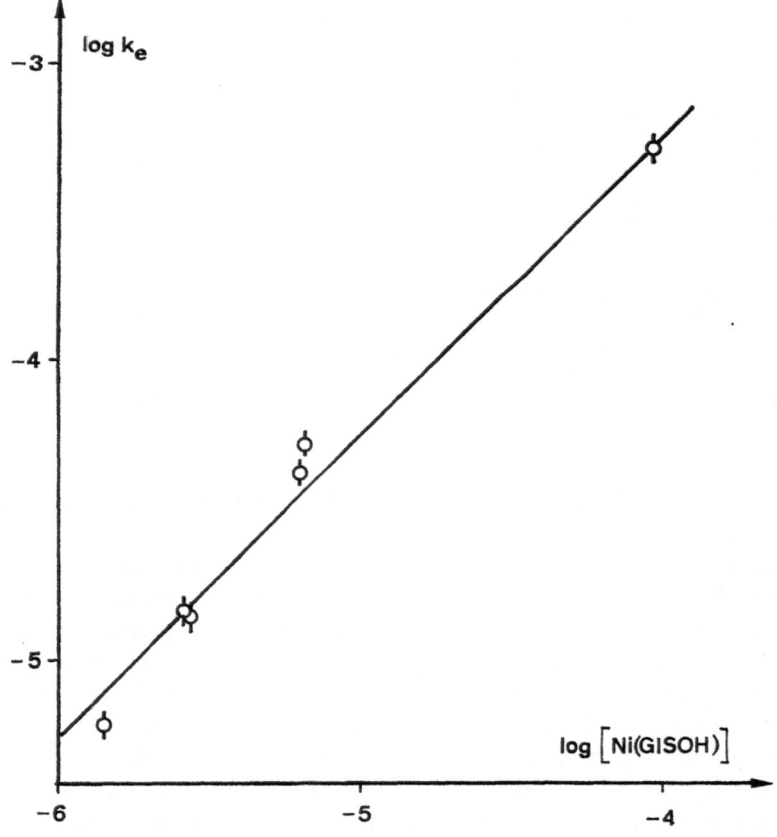

Fig. 18. The insertion of Ni(II) from glycylsarcosine complexes into TMPyP: k_e vs. $[\mathrm{Ni(GlS)(OH)}]$; 25 °C; $6.9 < \mathrm{pH} < 9.2$; $[\mathrm{Ni}]_t$ 0.9 to $4 \cdot 10^{-2}$ M; $[\mathrm{HGlS}]$ 2 to $8 \cdot 10^{-2}$ M

The reaction of $Ni(tir)^{2-}$ with TMPyP can be followed only over about one half-time because precipitates containing Ni(II), tir and TMPyP are formed. The second-order rate constant $k_2 \approx 0.4$ $M^{-1}s^{-1}$ (7) is about 10^{+4} times larger than for the insertion of Ni^{2+}, which agrees with the results obtained for Cu(II) under similar conditions.

3.5. Insertion of Fe(III) and Fe(II)

Apart from *Kassner* and *Wang's* (27) studies discussed in Section 2.3.1, no systematic work on Fe(II) has so far been reported. Iron porphyrins are usually prepared from Fe(II) compounds such as acetates. The variation of ligands is quite restricted since Λ may stabilize Fe(III) to such an extent that, for example, the reaction

$$Fe(II)\Lambda + H^+ \longrightarrow Fe(III)\Lambda + 1/2\ H_2 \tag{41}$$

could interfere with the insertion. It has been postulated that Fe(III) would quite generally form porphyrin complexes very slowly, possibly more so than Ni(II) (7).

Mononuclear hydroxocomplexes have been identified in studies on $Fe(edta)^-$ and Fe(nta), but dimerization to $(edta)FeOFe(edta)^{4-}$ is well established (53), and further oxygen-bridged binuclear complexes of Fe(III) have been characterized (54). Polynuclear hydroxocomplexes $Fe_n(OH)_{2.7n}^{0.3n+}$ (55) formed in chloride media at $1.8 < pH < 2.3$ do not react with TMPyP within one week, nor do mononuclear complexes of Fe(III) containing $CH_3CO_2^-$, N_3^-, $C_2O_4^{2-}$, HL^{2-} ($H_3L =$ Phosphoserine) or $(O_3P-CH_2-PO_3)^{4-}$.

Tiron ($H_2tir)^{2-}$ forms predominantly blue $Fe(tir)^-$, violet $Fe(tir)_2^{5-}$ and red $Fe(tir)_3^{9-}$ around pH 3, 5, and 7, respectively (56). As expected, the last complex is perfectly inert with respect to the incorporation reaction. However, the 1:1 and the 1:2 complexes containing four and two water molecules in the inner sphere react faster with TMPyP than hydrated Fe^{2+} (Table 9). An upper limit $k_2 \leq 10^{-7}$ $M^{-1}s^{-1}$ for the insertion of Fe^{3+} at 40 °C has been estimated (5). At the same temperature, the second-order rate constant for the reaction of $Fe(tir)^-$ with TMPyP has been determined to $2.2 \cdot 10^{-2}$ $M^{-1}s^{-1}$ (Table 9). Since Fe^{2+}

Table 9. The insertion of Fe(II) and Fe(III) into TMPyP in aqueous solution. H_2tir^{2-}: tiron (IX); k_2 ($M^{-1}sec^{-1}$); 40 °C. E_A: kcal/mol; 25°–40 °C; $\Delta S^{\pm}$ cal $\cdot mole^{-1}K^{-1}$

Complex	k_2	Medium	E_A, $\Delta S^{\pm}$
Fe^{2+}	$0.18 \cdot 10^{-2}$	$I = 0.5$ (NaCl) pH 3.2: CH_3CO_2(H,Na)	14.5 (± 1.5) -27 (± 5)
$Fe(tir)^-$	$2.2 \cdot 10^{-2}$	$I = 0.5$ (NaCl) pH 3.2: CH_3CO_2(H,Na)	17 (± 1) -17 (± 2.5)
$e(tir)_2^{5-}$	0.20	$0.1 < I < 0.6$ (NaCl, $NaCH_3CO_2$). pH 5: CH_3CO_2 (H,Na)	—

forms metalloporphyrin much more slowly than Fe(tir)$^-$, it is possible to follow the insertion

$$Fe(tir)^- + H_2P \longrightarrow Fe^{III}P + H_2tir^{2-} \tag{42}$$

in presence of Fe^{2+} at pH $\sim$3. The equilibration

$$Fe(III) + Fe(tir)^- \rightleftharpoons 2\,Fe^{2+} + C_6H_2O_2(SO_3)_2^{2-} \tag{43}$$

can be suppressed to a great extent by the addition of Fe^{2+}. Oxidation of tiron is no longer detectable when the concentration of Fe(Htir) vanishes in relation to Fe(tir)$^-$ and Fe(tir)$_2^{5-}$. It can therefore be ruled out that either of these complexes is able to incorporate iron in the chain (45, 46).

$$Fe(III) + Red \rightleftharpoons Fe(II) + Ox \tag{44}$$

$$Fe(II) + H_2P \longrightarrow Fe^{II}P + 2\,H^+ \tag{45}$$

$$Fe^{II}P + Fe(III) \rightleftharpoons Fe^{III}P + Fe(II) \tag{46}$$

$$Fe^{III}(TMPyP)^{5+}/Fe^{II}(TMPyP)^{4+}$$
$$E_0 = 0.18\,(\pm 0.01)\,V\,(vs.\,NHE) \tag{47}$$

Schoder (9) has determined the oxidation potential of the couple (47), and it can be deduced that in solutions containing Fe^{2+} and Fe(tir)$^-$ at pH 3, both Fe^{II}P and Fe^{III}P are formed. On the other hand, in 1 M HCl, the complexes FeCl$_n^{3-n}$ ($n = 0, 1, 2$) oxidize Fe^{II}(TMPyP) completely and rapidly to the extremely stable Fe^{III}(TMPyP)Cl^{4+}. Hence, the insertion of Fe(III) from intensely coloured tiron complexes can be stopped by quenching in acid solution whereby the tiron complexes are decomposed.

Any excess of tiron at pH $\sim$3 retards the reaction due to association of TMPyP($+4$) with H$_2$tir^{2-} (1:1, K $\sim$10^{+2}). When the pH is raised to 5, relative concentrations Fe(tir)$^- <$ Fe(tir)$_2^{5-} \gg$ Fe(tir)$_3^{9-}$ are established if the molar ratio tiron/Fe(III) equals two. Association of TMPyP($+4$) with Fe(tir)$_2^{5-}$ is even stronger than with H$_2$tir^{2-}.

$$Fe(tir)_2^{5-} + H_2P^{4+} \rightleftharpoons \{Fe(tir)_2 \cdot H_2P\}^- \quad K = 3 \cdot 10^{+3} \tag{48}$$

$$\{Fe(tir)_2 \cdot H_2P\}^- + 2\,HB \xrightarrow{k_z} Fe^{III}P^{5+} + 2\,H_2tir^{2-} + 2\,B^- \tag{49}$$

$$
\begin{aligned}
R &= k_2' \cdot [Fe(tir)_2^{5-}]^* \, [H_2P]^* \\
[Fe(tir)_2^{5-}]^* &= [Fe(tir)_2^{5-}] + [Fe(tir)_2 \cdot H_2P] \\
[H_2P]^* &= [H_2P] + [Fe(tir)_2 \cdot H_2P]
\end{aligned}
\tag{50}
$$

155

The coefficient k_2' (50) obtained from the experiment is the function

$$k_2' = \frac{K \cdot k_z}{(1 + [H_2P] \cdot K)(1 + [Fe(tir)_2] \cdot K)} \quad . \tag{51}$$

Fig. 19 shows data on solutions in which the association of H_2tir^{2-} with H_2P is negligible and $[Fe(tir)_3^{9-}]$ is exceedingly small.

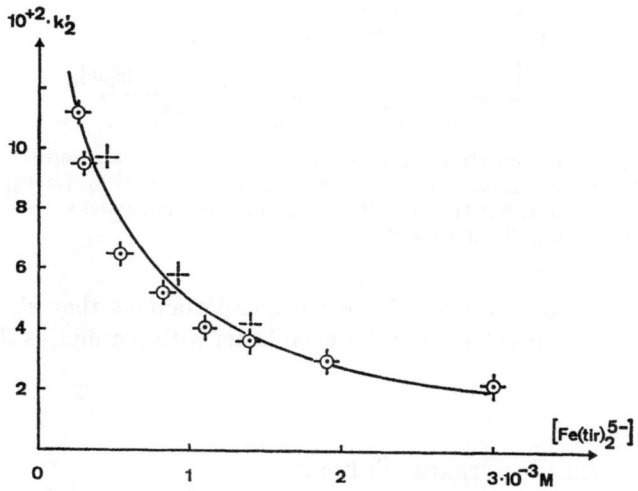

Fig. 19. Kinetic evidence for the association of TMPyP(+4) with $Fe(tir)_2^{5-}$. The coefficient k_2' is defined by Eq. (50). Molar ratio tir/Fe = 2; pH 5.1, $I = 0.5$ (NaCl, $CH_3CO_2^-$), 40 °C. The curve is calculated for $K = 3 \cdot 10^{+3}$; k_2 $(c \rightarrow 0) = 0.20$ $M^{-1}sec^{-1}$; $[P]_t \ll [Fe]_t$

The highest complex, $Fe(tir)_2^{9-}$, can no longer be neglected if at pH 5 more than two moles tiron per Fe(III) are introduced. Fig. 20 shows that association between $Fe(tir)_3^{9-}$ and TMPyP(+4) produces an inhibiting effect on the insertion. The k_2' coefficients are now extended functions of type (51). The association constant $K \approx 6 \cdot 10^{+5}$ accounts for this inhibiting effect. The rate coefficients summarized in Table 9 fit all data referring to a matrix of composition within the boundary concentrations

$$
\begin{array}{lll}
\text{Fe(III)} & 3 \cdot 10^{-4} \text{ to } 3 \cdot 10^{-3} \text{ M} \\
\text{tiron} & 1 \cdot 10^{-3} \text{ to } 1.1 \cdot 10^{-2} \text{ M} \\
\text{tiron/Fe(III)} & 1 \text{ to } 11 \\
\text{H}_2\text{P} & 2.4 \cdot 10^{-5} \text{ to } 3 \cdot 10^{-4} \\
\text{pH} & 3 \text{ to } 5.5 \\
\text{buffer} & \text{NaCH}_3\text{CO}_2/\text{CH}_3\text{CO}_2\text{H} \quad 0.05 \ldots \ldots 0.5 \\
\mu \text{ (NaCl, NaCH}_3\text{CO}_2) & 0.1 \text{ to } 0.6
\end{array}
\tag{52}
$$

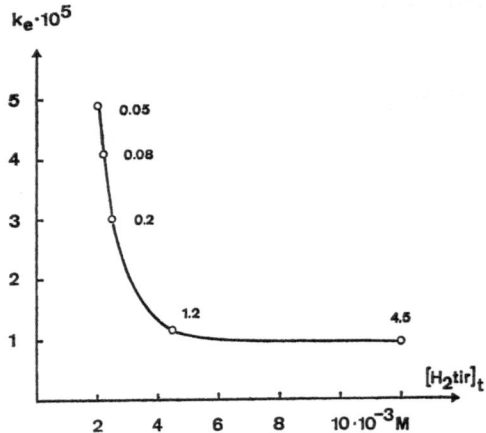

Fig. 20. Kinetic evidence for strong association of TMPyP($+4$) with Fe(tir)$_3^{9-}$. The solutions differ only in the total concentration of H$_2$tir^{2-}. [Fe]$_t = 1.0 \cdot 10^{-3}$ M, [H$_2$P]$_t = 2.4 \cdot 10^{-5}$ M; pH $= 5.0$; $I = 0.09$ (NaCl, NaCH$_3$CO$_2$); 40 °C. The numbers indicated are calculated concentrations of Fe(tir)$_3^{9-}$ in units of 10^{-5} M

The data collected in Table 9 are most important as they show that well-chosen complexes of Fe(III) may react even faster with porphyrins than hydrated Fe^{2+}.

4. Kinetic Studies in Organic Solvents

4.1. Bivalent Metal Ions in Acetic Acid

Insertion of Cu(II) and Zn(II) into TPyP was studied by *Choi* and *Fleischer* (57) in aqueous solution containing 8.3% acetic acid, and 0.01 M HClO$_4$ to ensure protonation of the peripheral pyridyl groups. It was found that the ratio $k_2^{Cu}/k_2^{Zn} \approx 30$ is nearly the same as for the parent TMPyP($+4$). When the fraction of acetic acid is gradually increased up to one, the relative rates vary as indicated in Fig. 21. The drop to the left largely results from increased electrostatic repulsion between Cu^{2+}, Cu(CH$_3$CO$_2$)$^+$ and H$_4$TPyP($+4$). This effect, of course, cannot be separated from the retardation by preferential solvation of the porphyrin with acetic acid. The maximum rate around 90% CH$_3$CO$_2$H undoubtedly stems from the gradual conversion of copper into Cu(CH$_3$CO$_2$)$_2$. The second drop results from the gradual formation of dimeric Cu$_2$(CH$_3$CO$_2$)$_4$,which is less reactive than the mononuclear complexes. On comparing relative rates for Mn(II), Co(II), Ni(II), Cu(II) and Zn(II), it is found that the order is modified as compared to aqueous media (Table 10). The solvated species in acetic acid are obviously no longer comparable with each other.

Brisbin's studies on the incorporation of these metals into haematoporphyrin- and protoporphyrin-IX-dimethylester also await better knowledge of species in pure acetic acid (58, 59). As for Cu(II), the rate law

$$R = k \cdot [\text{Cu(II)}]^{1/2} \cdot [\text{H}_2\text{P}] \tag{53}$$

W. Schneider

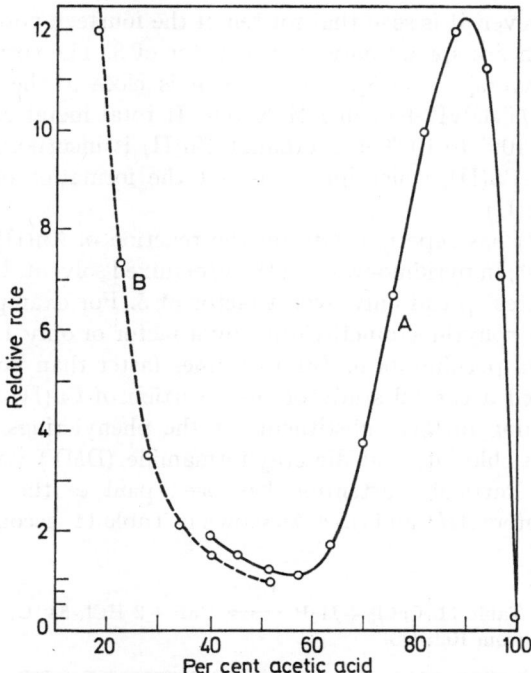

Fig. 21. The relative rates of the reaction of copper acetate with TPyP as a function of acetic acid content. Initial concentrations: Cu(II) $1.67 \cdot 10^{-2}$ M; curve A: TPyP $6.67 \cdot 10^{-5}$ M; curve B: TPyP $6.60 \cdot 10^{-5}$ M; 25 °C. From Ref. (57)

Table 10. Relative rates of formation of M(II)P under comparable conditions for each row

H$_2$P/Solvent	Fe(II)	Mn(II)	Co(II)	Ni(II)	Cu(II)	Zn(II)	Ref.
TMPyP/H$_2$O	$0.2 \cdot 10^{-3}$	$1.4 \cdot 10^{-3}$	$1.2 \cdot 10^{-3}$	$3 \cdot 10^{-5}$	1	$3 \cdot 10^{-2}$	(5, 11)
TPyP/CH$_3$CO$_2$H		$1.5 \cdot 10^{-2}$	0.2	$6 \cdot 10^{-2}$	1		(57)
Haemato/CH$_3$CO$_2$H	$3 \cdot 10^{-2}$		0.9		1		(59)

has been verified for H$_2$P = TPyP, diacetyldeutero- and haematoporphyrin. The spread induced by different substituents is rather small, k having the relative values 1, 2.3 and 4.1 for the three porphyrins (60).

4.2. Variation of the Porphyrin

Berezin (10) has recently reviewed the work of his group involving the reaction of cupric acetate and zinc acetate with fourteen different porphyrins[5]) including chlorins. As for ethanol solutions, crucial effects of complex equilibria are indicated

---

[5]) Porphin; Etio-I,II; Meso IX; Proto IX; Pyro XV; Phyllo XV; Pheophytin a, b; Pheophorbid a; Rhodin g$_7$; Chlorin e$_6$; synth. Chlorin; Etiochlorin II.

by the data. However, it is seen that for ten of the fourteen porphyrins, the rates of Cu(II) insertion are spread only over a factor of 3. The corresponding coefficients $k_e/[\mathrm{Cu}]_{tot}$ average to 1.5 $M^{-1}s^{-1}$, which is close to the value of about 2 for the pair $\mathrm{Cu}^{2+}/\mathrm{TMPyP}(+4)$ in 1 M KNO_3. If total metal concentrations are kept in the range 10^{-3} to 10^{-2} M in ethanol, Zn(II) is incorporated only slightly more slowly than Cu(II), which indicates that the formation of dimeric $M_2(ac)_4$ is restricted to Cu(II).

Hambright (61) has reported data on the reaction of Zn(II) with a series of twelve porphyrins[6]) in pyridine-water (11% v/v) mixed solvent. Under comparable conditions, the rates spread only over a factor of 5. For example, porphin itself differs from Uroporphyrinoctamethylester by a factor of only 1.5, whereas tetra-(p-methoxyphenyl)-porphin forms ZnP five times faster than porphin. *Longo* and *Adler* (43) reported a careful study of the insertion of Cu(II) into tetraphenyl-porphines containing further substituents at the phenyl rings. Cu(II) has been introduced as the chloride, and dimethylformamide (DMF) was shown to be a suitable solvent. Particular attention has been paid to the determination of activation parameters $\Delta H^{\pm}$ and $\Delta S^{\pm}$. As shown in Table 11, second-order rate con-

Table 11. $\mathrm{CuCl_2 + H_2P} \xrightarrow{\text{DMF}} \mathrm{CuP + 2\,HCl}$; 54 °C; from Ref. (43)

	$\Delta G^{\pm}$	$\Delta H^{\pm}$	$\Delta S^{\pm}$
TPP	19.6[b])	16.5	−10.0
p—Cl[a])	19.8	14.7	15.6
p—CH$_3$	19.5	12.1	−22.6
p—CN	20.3	12.0	−24.6
p—NO$_2$	20.2	13.5	−20.4
p—OH	18.8	8.5	−30.6
p—OCH$_3$	18.7	14.7	−12.2
o—OC$_2$H$_5$	20.2	23.8	11.2
m—CH$_3$	19.7	15.8	−11.9
o—CH$_3$	20.6	24.9	13.1
TPC	20.1	19.1	− 3.4
T(n—Pr)P	19.7	13.4	−19.2
Porphin	19.8	19.6	− 2,6
Etio I	19.4	16.6	− 8,6
Etio II	19.4	16.6	− 8,6

[a]) p—Cl = ms—tetra(p—chlorophenyl)porphin, etc.
[b]) Uncertainties in $\Delta G^{\pm} \sim \pm 0.03$, in $\Delta H^{\pm} \sim \pm 0.7$, in $\Delta S^{\pm} \sim \pm 2.0$.

---

[6]) Meso-IX-DME; Deutero-DME; Hemato-DME; Proto-DME; Dibromodeutero-DME; Porphin; Etio-III; Uro-I-OME; TPP; phenyl substituted TPP: p—OCH$_3$, p—CH$_3$; pCl. (DME: dimethylester; OME: octamethylester).

stants spread over a factor of 20. Taking porphin as a reference (k_2^0), the substituent effects to either side are rather small:

$$k_2 \,(\text{p—OCH}_3) = \frac{1}{3.4} \cdot k_2^0$$

$$k_2 \,(\text{o—CH}_3) = 5.4 \cdot k_2^0. \tag{54}$$

Phenyl substituents in meso positions do not affect the rate with respect to porphin. There is a considerable variation in ΔH^+ and ΔS^+, but the compensation effect is remarkable, as shown in Fig. 22.

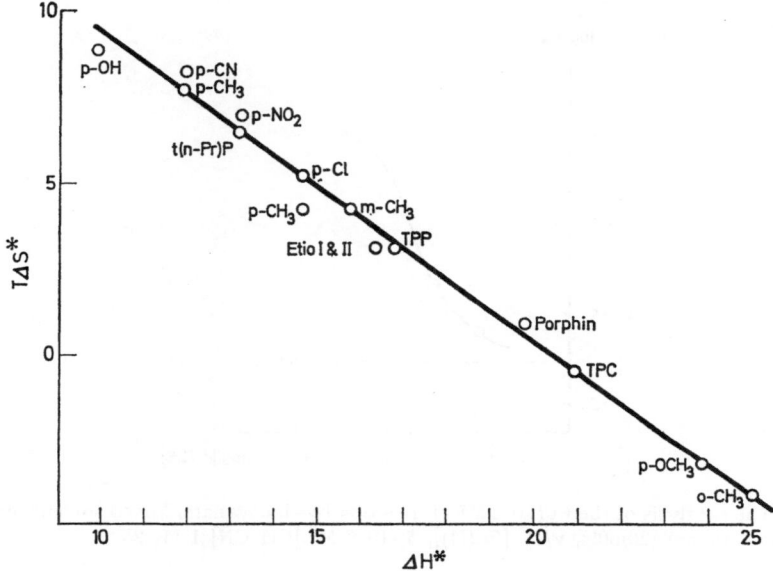

Fig. 22. The compensation effect for the reaction of various porphyrins (see Table 11) with CuCl$_2$ in dimethylformamide. From Ref. (43)

The parameter ΔH^+ is lowest for R = p-OH and highest for R = o—CH$_3$:

		ΔH^+	ΔS^+	
meso-C$_6$H$_4$—	{ o—CH$_3$	25	+13	(55)
	p—OH	8.5	−31 .	

This provides convincing evidence that the effect of these peripheral substituents is not electronically transmitted to the central part of the porphyrins. Rather solvation effects affect outer sphere association, $i.e.$ K_{os}, as discussed in Section 2.3.1, and $\varkappa$ in Eq. (13), where the electrostatic potential of peripheral dipoles is an important factor.

Altogether, the variation of porphyrin ligands does not provide as much insight into the relevant elementary steps as does the variation of Λ in MΛ.

4.3. Complexes of Cu(II), Fe(II) and Fe(III) in Dimethylsulphoxide-Ethanol

The mixed solvent dimethylsulphoxide-ethanol (20/80% v/v) provides better solubility of salts than pure DMSO. Metal ions select ligands according to the series $DMSO > H_2O > EtOH$. The solvated ion Cu^{2+} in the mixed solvent reacts with TMPyP($+4$) about five times more slowly toward the low ionic strength limit than does hydrated Cu^{2+} in aqueous solution (6). Whereas thiocyanate is oxidized by Cu^{2+} in water, $Cu(NCS)^+$ is a well-defined complex in the mixed solvent. The stability constant has been determined to $K = 8 (\pm 0.2)$ M^{-1} ($0.02 < I_c < 0.05$). Catalysis by NCS$^-$ is quite remarkable, as shown in Fig. 23.

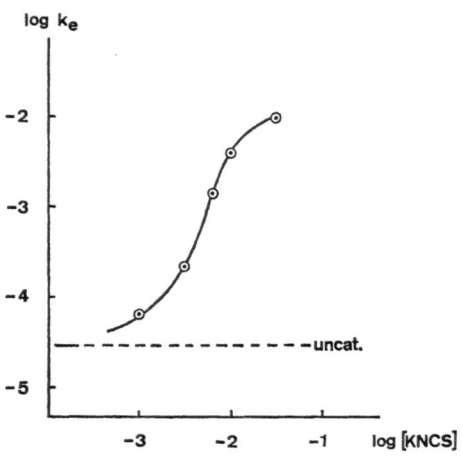

Fig. 23. The catalysis of the Cu(II)-TMPyP reaction by thiocyanate. Mixed solvent dimethyl-sulphoxide-ethanol (20/80% v/v). $[Cu(II)]_t$ 5 · 10^{-3} M; [CH$_3$CN] 1 M; 25 °C

The rate law

$$k_e = k_3 \cdot [Cu(NCS)^+] [NCS^-] \tag{56}$$

indicates either specific ion effects of NCS$^-$ or very reactive $Cu(NCS)_2$ having low stability, $K_2 \leq 1$. The same behaviour, including the rate dependence of type (56) and $k_3 \approx 10^{+3}$ is found for the catalysis by azide N$_3^-$ (6).

In DMSO/EtOH containing buffer (Him$^+$, im), Fe^{2+} has been found to react slightly faster than in aqueous solution, as expressed by $k_2 = 2.5 \cdot 10^{-3}$ vs. $0.6 \cdot 10^{-3}$. As expected, the complexes FeL$_n$Cl$_m$$^{3-m}$ (L = DMSO; $m \leq 2$) do not form Fe^{III}P at measurable rates. However, NCS$^-$ catalyzes the reaction sufficiently to permit rate studies. Lower thiocyanato-complexes of Fe(III) are stable over periods of several months in DMSO/EtOH, unlike their behaviour in water as carefully studied by *Dainton* (62). Since, in solutions containing Fe(NCS)$^{2+}$, the concentrations of Fe(II) are negligible compared to Fe(III), insertion via Fe(II) can be ruled out as a significant contribution to the rates observed. The data in Fig. 24 fit a rate law of the type (56), $k_3 \approx 0.3$ M^{-2}s^{-1}.

Table 12. Metalloporphyrin formation in the mixed solvent dimethylsulphoxide-ethanol (20/80% v/v). TMPyP

Complex	k_2	k_3	Conditions	Temp., °C
$Cu^{2+}_{solv.}$	0.05		$Cu(Ts)_2 \leq 5 \cdot 10^{-4}$ M. Veronal buffer 10^{-3} M. CH_3CN 1 M	25
$Cu(NCS)^+$ $Cu(N_3)^+$		$10^{+3} =$	$k_e/[CuL^+][L^-]$. Veronal buffer 10^{-3} M. CH_3CN 1 M.	25
Fe^{2+}	$0.25 \cdot 10^{-2}$ $1.1 \cdot 10^{-2}$		Imidazol buffer: Him^+, im, Cl^- $2.5 \cdot 10^{-2}$ M	25 40
$Fe(NCS)^{2+}$		$0.30 =$	$k_e/[Fe(NCS)^{2+}][NCS^-]$; Him^+, im 2.5 to $5 \cdot 10^{-3}$ M	40

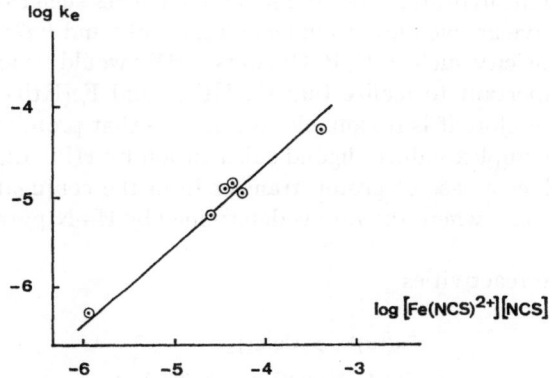

Fig. 24. The insertion of Fe(III) into TMPyP in the mixed solvent DMSO/EtOH. Imidazole buffer; 40 °C

5. Conclusions

5.1. Mechanisms

It became evident during the discussion in Sections 3 and 4 that the inner sphere Λ of complexes $M\Lambda$ determines the rates of metalloporphyrin formation to a high degree. Moreover, the conceptual separation of outer-sphere association from specific-site association is supported by the effects of solute components such as Ts^-, H_2tir^{2-}, $Fe(tir)_2^{5-}$, $Fe(tir)_3^{9-}$, which merely represent the most obvious cases with regard to association phenomena. It is justified to factorize a second-order rate constant according to

$$k_2 = K_\sigma \cdot k\,(s, p, d) \tag{57}$$

where K_σ accounts for all equilibrium-type processes which proceed much faster than the real incorporation steps. It may be asked whether further factorization is possible of k (s, p, d), involving substitution (s) at $M\Lambda$ by N(pyrrole), proton abstraction (p) from H_2P, and deformation (d) of H_2P. Kinetic evidence has been presented which stresses the importance of out-of-plane deformation.

In N-methylated porphyrins, the planarity of the porphyrin skeleton is lost and the basicity increases considerably (63). Rates of incorporation of Zn^{2+} and Cd^{2+} are higher by five orders of magnitude for N-methyletioporphyrin as compared to etioporphyrin (64)[7]. Dynamic distortions most likely involve smaller amplitudes than the static ones induced by N-methylation. If the relevant frequencies are larger than the exchange frequencies of the most labile ligands in $M\Lambda$ (i.e. solvent molecules), the last two factors (s, p) obviously have to be considered in the context of a concerted mechanism.

In aqueous media, the exchange rates of water in $M\Lambda$ of all complexes considered in Section 3 are considerably higher than the rate coefficients k_z. Hence, it is unlikely that further labilizing of H_2O at $M\Lambda$ will enhance rates beyond those of the parent hydrated ions. Moreover, complexes such as $Cu(tir)^{2-}$, $Fe(tir)^-$, and $Fe(tir)_2^{5-}$ have an even lower tendency than Cu^{2+} and Fe^{3+} to bind a nitrogen ligand of low basicity such as H_2P. Of course, HP^- would be an extremely strong donor. It is important to realize that $Cu(Htir)^-$ and $Fe(Htir)$ are well-identified species (56). Therefore it is reasonable to conclude that proton transfer from H_2P to tiron in the complex induces ligand substitution by HP^-. Clearly, the situation described involves a case of proton transfer from the conjugate acid of a strong base to a weak base where the rate is determined by H—N(pyrrole) bond breaking (65).

The relative reactivities

$$Cu(tir)^{2-} > Cu(ac)_2$$
$$Ni(GG_-H)\ (OH)^- > Ni(GGH_-) \tag{58}$$
$$Cu(dip)OH^+ > Cu(dip)^{2+}$$

are far more obvious than the catalytic effects of N_3^- and NCS^- demonstrated for Cu(II) and Fe(III). The weaker effect of NH_3 as compared to N_3^- is an important result. Although NH_3 is the stronger base, it can be protonated only when removed from the complex, opposite to OH^- in $Cu(OH)^+$. Monodentate azide in $Cu(N_3)^+$ certainly has rather low end-on basicity, but may assist in proton transfer. Apart from quantitative differences, it is a property all catalytically effective ligands have in common that they can be protonated in the complex without extensive predissociation. Hence, in a qualitative sense, ligand catalysis can be rationalized in terms of effects on K_σ, and on proton transfer from H_2P to Λ. An obvious case of effects on K_σ is provided by the data on the insertion of Fe(III) from $Fe(tir)^-$ and $Fe(tir)_2^{5-}$ into TMPyP (Table 9). Whereas the second-order rate constant k_2, i.e. the product $K_0 k_z$ according to Eqs. (11) and (12), is larger for $Fe(tir)_2^{5-}$, the first-order parameter k_z is larger for $Fe(tir)^-$ (Table 13).

---

[7]) Solvent dimethylformamide; metal chlorides; low ionic strength, 25 °C.

Table 13. The formation of $Fe^{III}(TMPyP)$ from $Fe(tir)_{1,2}^{-,5-} \cdot k_2 = k_z \cdot K$.
K: Stability constant of $\{Fe(tir)_n \cdot TMPyP\}$

	k_2	K	k_z, sec^{-1}
$Fe(tir)^-$	$2.2 \cdot 10^{-2}$	$3 \leq K \leq 10$	$\sim 5 \cdot 10^{-3}$
$Fe(tir)_2^{5-}$	0.2	$3 \cdot 10^{+3}$	$\sim 7 \cdot 10^{-5}$
$Fe(tir)_3^{9-}$	negl.	$6 \cdot 10^{+5}$	negl.

Certainly, steric effects due to the larger size of $Fe(tir)_2^{5-}$ are involved in the lowering of k_z. The order of magnitude $k_z \sim 5 \cdot 10^{-3}$ sec^{-1} referring to $Fe(tir)^-$ is an important result because the rate of dissociation.

$$Fe(tir)^- + H^+ \longrightarrow Fe^{3+} + Htir^{3-} \qquad (59)$$

is estimated to be much smaller.

Hence, dissociation of tiron occurs in a concerted fashion when the porphyrin is incorporated into the first coordination sphere. Obviously, if proton transfer from H_2P has been established, the series of substitutions occurring at $M\Lambda$ are rather fast steps dominated by kinetic chelate effects.

5.2. Experimental Studies

Not much attention has been paid in this review to activation parameters derived from the temperature dependence of second-order rate constants. From available data the crude rule emerges that, in the reaction of $M\Lambda$ with a porphyrin, the Arrhenius parameter E_A is no more than some 4 to 6 kcal higher than for normal substitution reactions of the same $M\Lambda$. However, there is so far no system in which independent studies have been made of the association of the reactants, including their temperature dependence. Similarly, rates of deprotonation of porphyrins have not yet been determined for obvious reasons. Clearly, a better understanding of quantities would emerge if first-order rate constants of type k_z could be derived from second-order coefficients k_2 in a reliable way, and if activation parameters of proton transfer from H_2P to suitable bases could be compared with the same parameters obtained from k_z. Turning to applications suggested by data reported in Section 3, it may be useful in certain cases to detect mononuclear hydroxocomplexes via enhanced rates of metalloporphyrin formation.

6. References

1. *Falk, J. E.*: Porphyrins and metalloporphyrins. Amsterdam: Elsevier 1964.
2. *Lowe, M. B., Phillips, J. N.*: Nature *190*, 262 (1961); *194*, 1058 (1962).
3. *Hambright, P.*: Coord. Chem. Rev. *6*, 247 (1971).
4. *Schneider, W.*: Proc. XV. ICCC 392 (1973).
5. *Schneider, W., Kläntschi, K., Stern, Ch., Blöchlinger, R., Schluep, H., Schoder, A.*: unpublished work.

6. *Stern, Ch.:* Dissertation ETH-Z. Nr. 5029 (1973).
7. *Blöchlinger, R.:* Dissertation ETH-Z. Nr. 5318 (1974).
8. *Schluep, H. P.:* Dissertation ETH-Z. Nr. 5364 (1974).
9. *Schoder, A.:* Dissertation ETH-Z. Nr. 5491 (1975).
10. *Berezin, B. D.:* Usp. Khim. *42*, 2007 (1973), and references quoted therein.
11. *Hambright, P., Chock, P. B.:* J. Am. Chem. Soc. *96*, 3123 (1974).
12. *Fleischer, E. B.:* Accounts Chem. Res. *3*, 105 (1970).
13. *Fuhrhop, J. H.:* Struct. Bonding *18*, 1 (1974).
14. *Ostfeld, D., Tsutsui, M.:* Accounts Chem. Res. *7*, 52 (1974).
15. *McHugh, A. J., Gouterman, M., Weiss, Ch. Jr.:* Theoret. Chim. Acta (Berl.) *24*, 346 (1972), and references quoted therein.
16. *Cabiness, D. K., Margerum, D. W.:* J. Am. Chem. Soc. *92*, 2151 (1970).
17. *Pasternack, R. F.:* Ann. N. Y. Acad. Sci. *206*, 616 (1973).
18. *Caughey, W. S., Eberspaecher, H., Fuchsman, W. H., McCoy, S.:* Ann. N. Y. Acad. Sci. *153*, 722 (1969).
19. *Fleischer, E. B., Palmer, J. M., Srivastava, T. S., Chatterjee, A.:* J. Am. Chem. Soc. *93*, 3162 (1971).
20. *Stein, T. P., Plane, R. A.:* J. Am. Chem. Soc. *91*, 607 (1969).
21. *Mauzerall, D.:* Biochemistry *4*, 1801 (1965).
22. *Storm, C. B., Teklu, Y.:* J. Am. Chem. Soc. *94*, 1745 (1972).
23. *Storm, C. B., Teklu, Y., Sokoloski, E. Y.:* Ann. N. Y. Acad. Sci. *206*, 631 (1973).
24. *Macquet, J. P., Theophanides, T.:* Can. J. Chem. *51*, 219 (1973).
25. Ref. (*7*), p. 176.
26. Ref. (*7*), p. 28.
27. *Kassner, R. J., Wang, J. H.:* J. Am. Chem. Soc. *88*, 5170 (1966).
28. Ref. (*7*), p. 27.
29. *Eigen, M., Kruse, W., Maass, G., De Maeyer, L.:* In: Progress in reaction kinetics, Vol. 2., p. 286. London: Pergamon Press 1964.
30. *Hewkin, D. J., Prince, R. H.:* Coord. Chem. Rev. *5*, 45 (1970), and references quoted therein.
31. *Fuoss, R. M.:* J. Am. Chem. Soc. *80*, 5059 (1958).
32. *Bjerrum, J.:* In: Proc. 3. Symp. Coord. Chem., Vol. 2, p. 9. Budapest: Akadémiai Kiado 1971.
33. *Robinson, R. A.: Stokes, R. H.:* Electrolyte solutions. London: Butterworths 1959.
34. *Cheung, S. K., Dixon, F. L., Fleischer, E. B., Jeter, D. Y., Krishnamurthy, M.:* Bioinorg. Chem. *2*, 281 (1973).
35. *Baker, H., Hambright, P., Wagner, L., Ross, L.:* Inorg. Chem. *12*, 2200 (1973).
36. *Das, R. R.:* J. Inorg. Nucl. Chem. *34*, 1263 (1972).
37. *Diebler, H.:* In: Proc. 3. Symp. Coord. Chem., Vol. 2, p. 53. Budapest: Akadémiai Kiado 1971.
38. *Diebler, H.:* Ber. Bunsenges. Physik. Chem. *74*, 268 (1970).
39. *Bernauer, K., Fallab, S.:* Helv. Chim. Acta *44*, 1287 (1961); *45*, 2487 (1962).
40. *Schiller, I., Bernauer, K., Fallab, S.:* Experientia *17*, 540 (1961). — *Schiller, I., Bernauer, K.:* Helv. Chim. Acta *46*, 3002 (1963).
41. *Sillén, L. G., Martell, A. E.:* Chem. Soc. Spec. Publ. No 17 (1964); No. 25 (1971).
42. *Lowe, M. B., Phillips, J. N.:* Proc. 11 ICCC 16 (1968).
43. *Longo, F. R., Brown, E. M., Quimby, D. J., Adler, A. D., Meot-Ner, M.,* Ann. N. Y. Acad. Sci. *206*, 420 (1973).
44. *Baker, H., Hambright, P., Wagner, L.:* J. Am. Chem. Soc. *95*, 5942 (1973).
45. *Schindler, P.:* private communication.
46. *Schneider, W., Schoder, A., Anner, Ch., Berner, R.:* unpublished work.
47. *Kakihana, H., Amaya, T., Maeda, M.:* Bull. Chem. Soc. Japan *43*, 3155 (1970).
48. *Wenger, H.:* Dissertation ETH-Z Nr. 3446 (1964).
49. *Hemmerich, P., Sigwart, Ch.:* Experientia *19*, 488 (1963).
50. *Anderegg, G., Blöchlinger, R.:* unpublished work.
51. *Martin, R. B., Chamberlin, M., Edsall, J. T.:* J. Am. Chem. Soc. *82*, 495 (1960).

52. *Kim, M. K., Martell, A. E.:* J. Am. Chem. Soc. *89,* 5138 (1967.) — *Davies, G., Kustin, K., Pasternack, R. F.:* Inorg. Chem. *8,* 1535 (1969). — *Lim, M. C., Nancollas, G. H.:* Inorg. Chem. *10,* 1957 (1971). — *Pasternack, R. F., Gipp, L., Sigel, H.:* J. Am. Chem. Soc. *94,* 8031 (1972).
53. *Schugar, H., Walling, C., Jones, R. B., Gray, H. B.:* J. Am. Chem. Soc. *89,* 3712 (1967).
54. *Murray, K. S.:* Coord. Chem. Rev. *12,* 1 (1974).
55. *Schneider, W., Anner, Ch.:* unpublished work.
56. *Schwarzenbach, G., Willi, A.:* Helv. Chim. Acta *34,* 528 (1951).
57. *Choi, E. I., Fleischer, E. B.:* Inorg. Chem. *2,* 94 (1963).
58. *Kingham, D. J., Brisbin, D. A.:* Inorg. Chem. *9,* 2034 (1970).
59. *Brisbin, D. A., Richards, G. D.:* Inorg. Chem. *11,* 2849 (1972).
60. *James, J., Hambright, P.:* Inorg. Chem. *12,* 474 (1973).
61. *Hambright, P.:* Ann. N. Y. Acad. Sci. *206,* 443 (1973).
62. *Betts, R. H., Dainton, F. S.:* J. Am. Chem. Soc. *75,* 5421 (1953).
63. *Jackson, A. H., Dearden, G. R.:* Ann. N. Y. Acad. Sci. *206,* 151 (1973).
64. *Shah, B., Shears, B., Hambright, P.:* Inorg. Chem. *10,* 1828 (1971).
65. *Eigen, M.:* Pure Appl. Chem. *6,* 97 (1963).

Received January 7, 1975

Hydrogen-Deuterium Exchange in Aromatic Compounds

Milton Orchin and **Donald M. Bollinger**

Department of Chemistry, University of Cincinnati
Cincinnati, Ohio 45221, U.S.A.

Table of Contents

I. Introduction

Ever since D_2O became reasonably accessible in the early 1930's, it was inevitable that there would be a demand for deuterium labelled compounds and accordingly hydrogen-deuterium exchange reactions would become commonplace. The interest in deuterium labelled compounds rests on their demonstrated use as powerful tools for elucidating the detailed mechanisms of many organic reactions. Two broad approaches are used in such studies. The fate of the deuterium atom is followed during the course of a reaction and often gives valuable clues as to mechanism. Thus the nature of the NIH shift (1) (named after the National Institutes of Health in Washington, D. C. where the discoverers were employed) involving either the in *vitro* or the in *vivo* rearrangement of arene oxides to phenols was demonstrated by the following reaction (2):

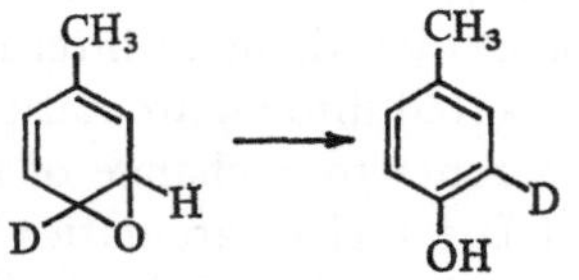

This rearrangement proceeded with 85 percent migration and retention of D and thus was unlikely to have involved a cyclohexanone intermediate. The second broad application of D-labelled compounds in mechanistic studies consists of experiments in which the rate of a reaction involving the breaking of a C—H bond is compared to the rate of the parallel reaction for the C—D compound. If a kinetic isotope effect is found, then the breaking of the C—H bond is more or less implicated in the rate-determining step of the reaction.

Deuterium (and tritium) labelled compounds are not only useful for mechanistic studies but valuable information concerning both ground state and excited state properties of molecules has been obtained by the use of compounds containing isotopes of hydrogen.

In this review paper we will be primarily concerned with the exchange between hydrogen and deuterium in aromatic compounds. Completely deuterated aromatics are available commercially. These are usually prepared by repeated treatment of the aromatic compound with D_2O at about 350° in an autoclave in the presence of a platinum catalyst (PtO_2 reduced with D_2)[1]. Such heterogeneous exchange reactions are not included in this review.

Because aromatic compounds are subject to electrophilic attack, D^+ provides a ready entry of deuterium into the aromatic molecule. The reaction of various anhydrides with D_2O is one of the most common methods for providing ^{2}H protic acids. The acidity of protic acids can be enhanced by "co-catalysts", which are usually Lewis acids. Hence electrophilic substitutions by protic acids and Lewis acids are of primary interest in hydrogen isotope exchange.

Exchange reactions of aromatic compounds catalyzed by base are also well-documented. The generation of benzyne intermediates by loss of HX (or DX) from haloaromatics in the presence of a strong base such as NH_2^- (or ND_2^-) is responsible for H—D exchange. Complexation of aromatics to zero-valent transition metals to give complexes such as π-$C_6H_6Cr(CO)_3$ enhances the acidity of the protons of the complexed aromatic and permits base-catalyzed exchange.

The addition of molecular hydrogen to unsaturated compounds catalyzed by Group VIII metals was first investigated at the beginning of this century by *Sabatier* and *Senderens* (3) who conducted vapor-phase hydrogenations over a nickel catalyst. Despite the fact that the reaction has been known for three-quarters of a century and despite the armamentarium of very sophisticated instrumentation for studying reactions on surfaces, the mechanism of heterogeneous hydrogenation is still not completely understood.

---

[1] Private communication from *E. Stohler* and *A. Bader*.

During the last twenty years, the development of organometal chemistry and the discovery of many soluble transition metal catalysts which promote the hydrogenation of organic compounds (4) have aided our understanding of the function of the catalyst. The formation of hydridometal intermediates in which the central metal atom of a complex is inserted into an aromatic C—H bond has been clearly demonstrated and provides a route to exchange of *ortho* hydrogens in aromatic phosphines. The exchange of D for H in aromatics catalyzed by Pt(II) salts in CH_3CO_2D proceeds by a similar insertion of the Pt into a C—H bond.

This review is not an exhaustive or complete literature survey of all exchange reactions; rather it is an attempt to indicate the variety and scope of reactions which have been used for this purpose. Discussions of mechanism are included in the hope that such discussions will reveal common pathways for breaking and making C—H bonds in seemingly unrelated reactions. Practically all the discussion which follows is concerned with ^1H and ^2H isotopes although, except for the magnitude of rates and the special techniques for handling, the principles employed for exchange apply equally well to ^3H (tritium).

II. Acid-Catalyzed Exchange

A. Introduction

Electrophilic substitution is the classical reaction which is always cited as characterizing aromatic compounds as distinguished from olefinic compounds. Among the many electrophilic substitution reactions, the Friedel-Crafts synthesis has played a central role in both practical and theoretical organic chemistry. The substitution of one hydrogen isotope by another, *e.g.*, the substitution of D^+ for H^+, is a specific case of conventional aromatic substitution.

A variety of acidic systems has been used to catalyze isotopic hydrogen exchange in aromatic hydrocarbons. A list of such systems appears elsewhere (5); it includes inorganic acids, organic acids, and mixtures of the two with or without the addition of Lewis acids such as metal halides or salts which are frequently used as Friedel-Crafts catalysts.

B. The Mechanism of Electrophilic Substitution

Characteristically, electrophilic substitution, which was once thought to proceed by a rather simple mechanism, turns out to be quite a complex reaction. A key development in understanding mechanistic pathways has been the studies involving kinetic isotope effects. Much of this work indicated a two-step reaction process with the implied formation of an intermediate; but the use of isotopes has also added additional complexities to the interpretation of electrophilic substitution. In addition to the kinetic isotope effect, studies on the formation, stability, structure, and reactions of sigma and pi complexes have also shed light on the role of intermediates in the substitution process.

1. The Kinetic Isotope Effect. The substitution of a heavier isotope for a lighter one results in a lowering of the zero-point energy of the bond to which the heavier isotope is attached. Replacement of hydrogen by deuterium in the C—H bond of an organic molecule strengthens the bond by about 1.2 kcal/mole at room temperature; accordingly more energy is required to break the C—D bond than is required to break the corresponding C—H bond. This greater energy requirement results in a decreased reaction rate for the deuterated compound when the breaking of the C—D bond is wholly or partially rate-determining. This rate depression is known as the kinetic isotope effect.

Consider a generalized electrophilic substitution reaction in which E^+ is any electrophile:

$$ArH + E^+ \underset{k_{-1}}{\overset{k_1}{\rightleftharpoons}} ArHE^+ \overset{k_2}{\longrightarrow} ArE + H^+ \tag{1}$$

The potential energy-reaction coordinate diagram for this reaction when $k_2 \ll k_{-1}$ is shown in Fig. 1. The greater energy required to break the C—D bond in the second transition state slows the rate of substitution on the deuterated compound because of the higher energy of activation. However, if the step leading to the first transition state were rate-determining, then no primary kinetic isotope effect would be observed. In this case $k_2 \gg k_{-1}$. Nitration reactions are the only class of reactions which clearly fall in this later category. Kinetic isotope effects have been observed for all other types of electrophilic aromatic substitutions. However the magnitude of these effects has not always been indicative of a primary kinetic isotope effect.

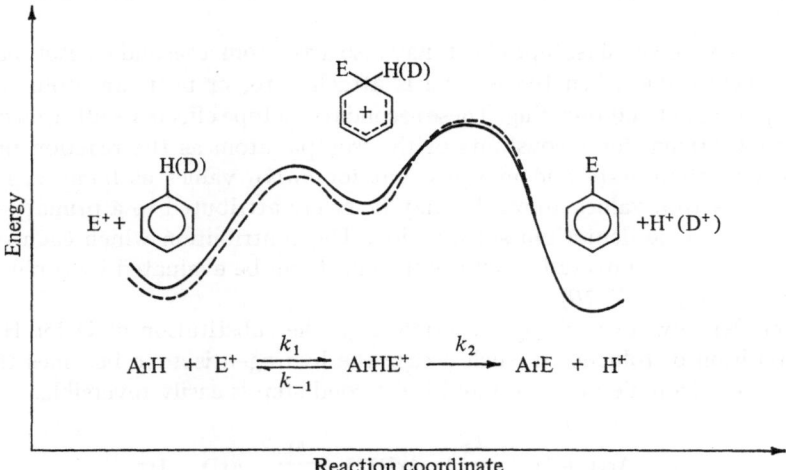

Fig. 1. Potential energy profile showing the primary kinetic isotope effect in a two-step reaction. The dashed curve indicates the potential energy profile for the deuteriobenzene

If all the zero-point energy is lost in the stretching of the C—H bond in the transition state, then the calculated primary isotope effect should be around 7 at room temperature (6). This means the isotope rate ratio, k_H/k_D, where k_H is the rate constant for substitution of protium and k_D is the rate constant for substitution of deuterium, should be about 7. Several factors may contribute to the lowering of this isotope rate ratio. A reaction in which $k_{-1} \simeq k_2$ will give a smaller isotope effect because the second step is not completely rate-determining. The observed isotope effect will depend upon the relative values of k_2 and k_{-1} [cf. Eq. (1)]. These rate constants control the partitioning of the intermediate between product and starting material.

The overall observed rate constant of the two-step reaction will have the general form of Eq. (2) (7):

$$k_{obsd.} = \frac{k_1 k_2}{k_{-1} + k_2} = \frac{k_1 \dfrac{k_2}{k_{-1}}}{1 + \dfrac{k_2}{k_{-1}}} \tag{2}$$

For the case shown in Fig. 1, k_1 for $C_6H_6 + E^+$ equals k_1 for $C_6H_5D + E^+$ and the values for k_{-1} for both reactions are also equal. In this situation k_{obsd} will depend almost exclusively on k_2 and kinetic isotope effects will be observed. When $k_{-1} \ll k_2$, $k_{obsd} \simeq k_1$ and no isotope effect occurs. However, for many reactions, and particularly for hydrogen exchange reactions, the values of k_{-1} and k_2 approach each other and the above simplification of Eq. (2) cannot be made. In such situations isotope effects smaller than 7 may be observed.

Residual zero-point energy in the transition state will also decrease the observed isotope effect. This is believed to occur when the conjugate base, to which the hydrogen transfers, participates in the transition state of a rate-determining second step, k_2 (8).

An observed small isotope effect may also arise from a secondary isotope effect. Such effects occur when the isotope is attached to, or near, an atom which is participating in bond breaking. These secondary isotope effects result from changes in the vibrational force constants of the isotopic atom as the reaction proceeds to the transition state and may account for k_H/k_D values as large as 1.74 (9). Thus only k_H/k_D values above 2.0 may be safely attributed to a primary isotope effect for a single deuterium substitution. The contribution which each of these effects makes to an overall small isotope effect can be evaluated by sophisticated kinetic techniques (7, 10).

Consider now an exchange reaction, e.g., the substitution of D for H in the *para* position of toluene. In such a case the hydrogen isotope becomes the substituting electrophile in Eq. (1) and the second step is easily reversible.

$$\text{ArH} + \text{D}^+ \underset{k_{-1}}{\overset{k_1}{\rightleftharpoons}} \text{ArHD}^+ \underset{k_{-2}}{\overset{k_2}{\rightleftharpoons}} \text{ArD} + \text{H}^+ \tag{3}$$

The energy-reaction coordinate diagram for such reactions is shown in Fig. 2. For the exchange reaction, the k_H/k_D values cannot be interpreted as simply as before. In Fig. 1, the energy of the first transition state is essentially independent of whether the protio aromatic or the deuterio aromatic is reacting with E^+; neither k_1 nor k_{-1} is affected by substitution of D for H. Accordingly, any decrease in observed rates for the deuterated compound must come from an increased

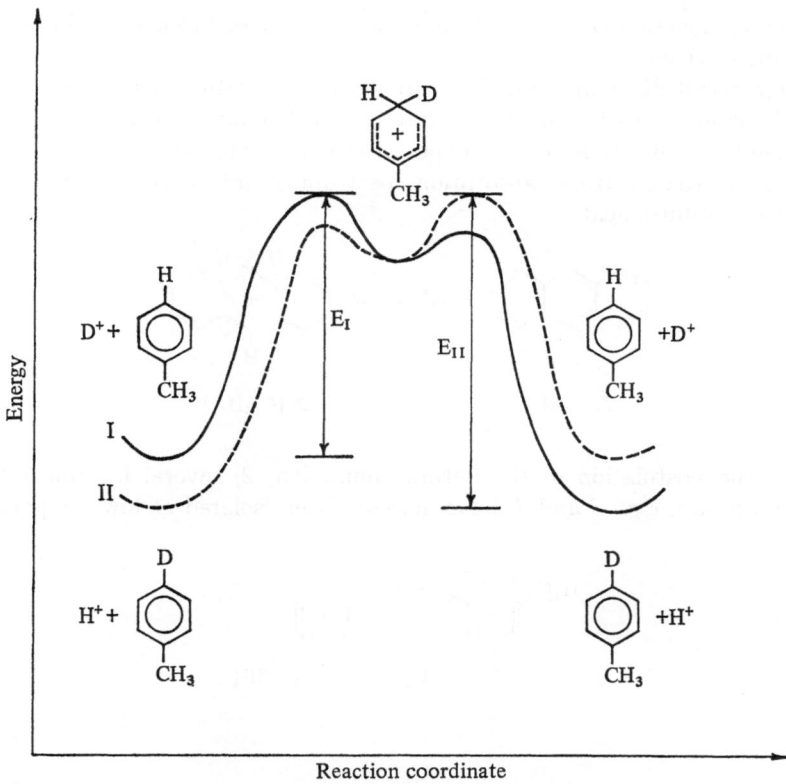

Fig. 2. Potential energy profile: (I) for H—D exchange; (II) for D—H exchange

energy of activation for the second step, k_2. For the exchange reactions shown in Fig. 2, the electrophile is different for reaction I (deuteriodeprotonation) and reaction II (protiodedeuteration) but both proceed through a common intermediate. Hence k_1 as well as k_{-1}, are not the same for the two reactions, and kinetic isotope effects cannot be effectively isolated on one particular transition state. As Fig. 2 implies, any difference in observed rates is probably due to the decreased zero-point energy of the deuterium substituted toluene (11), i.e., an increased activation energy for dedeuteration, k_D.

Kinetic isotope effects in the exchange reaction can, however, be evaluated by the use of a tritium label. Thus the ratio of the rates for protiodedeuteration ($H^+ + C_6H_5D$), k_D, and protiodetritiation ($H^+ + C_6H_5T$), k_T, may be determined and these k_D/k_T values may be related to k_H/k_D values by known equations (12, 13). In 81 percent H_2SO_4, the k_D/k_T value is 1.5 for a single position of benzene, and 1.5, 1.7, and 1.9 for the meta, ortho, and para positions of toluene respectively (14).

In hydrogen exchange reactions, the isotope effect apparently increases for substitution at the more reactive positions, whereas in most other electrophilic substitution reactions the isotope effect decreases. This reversal has been ascribed to the importance of a methylene rocking vibration in the transition state of the exchange reaction (15); methylene groups cannot be formed in other electrophilic substitutions.

2. Sigma and Pi Complexes. The cationic species with a delocalized positive charge is assumed to be an intermediate in electrophilic substitutions, Fig. 1. Experimental verification of such a species was first claimed (16) on the basis of a red shift in the electronic absorption spectrum of anthracene, 1, when it was dissolved in sulfuric acid:

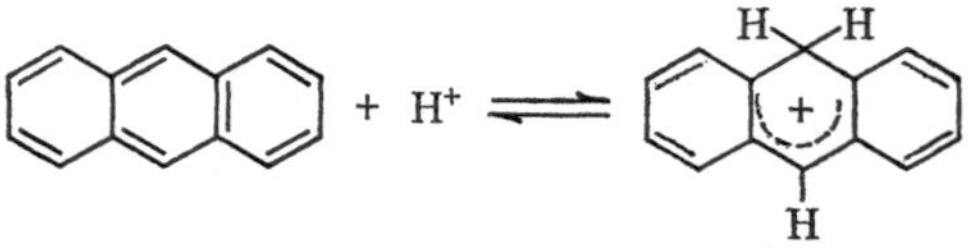

$$+ H^+ \rightleftharpoons \qquad\qquad\qquad (4)$$

1, $C_{14}H_{10}$ *2*, $[C_{14}H_{11}]^+$

Since the postulation of the anthracenium ion, *2*, several intermediates of similar structure, *e.g.*, *3* and *4*, have actually been isolated at low temperatures (17):

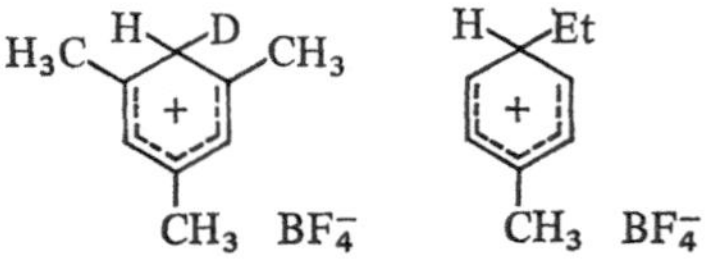

3   *4*

Upon warming, the intermediate *3* yields about 40 percent mesitylene and 60 percent mesitylene-d_1 by loss of a deuteron and proton, respectively. This result is consistent with an increased energy barrier for C—D bond cleavage in accordance with Fig. 2.

Species such as *2*, *3*, and *4* are called sigma complexes because the entering proton (or electrophile) is bonded to a specific carbon atom of the aromatic ring by a sigma bond. Such compounds are colored, are conducting at their melting points, and are only slighly soluble in organic solvents.

In contrast to the behavior of aromatics with H_2SO_4 and HBF_4, when $HCl_{(g)}$ is dissolved in an aromatic compound the solution is colorless and non-conducting. Solubility measurements indicate that a 1:1 complex is formed which, with benzene, may be written as *5*:

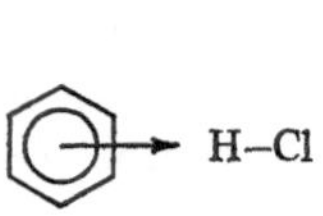

H—Cl

5

If DCl rather than HCl is employed, no exchange with aromatic hydrogens is observed. Complexes such as *5* are called charge transfer, or π, complexes. The H atom originally belonging to HCl is not associated specifically with any particular carbon atom but can be visualized as being held in the π cloud of the aromatic system and only partially separated from Cl. In such complexes, the HCl bond is more highly polarized than in the free state; some charge has been transferred.

The π complexes were first ascribed kinetic importance in exchange reactions after it was observed (*18*) that the logarithm of the dedeuteration rate for several aromatic compounds was linearly dependent on the acid strength of the protic acid (measured by the Hammett acidity function, H_0, which will be discussed later). This was interpreted as requiring the separation of A$^-$ from the aromatic system before the slow step, k_2:

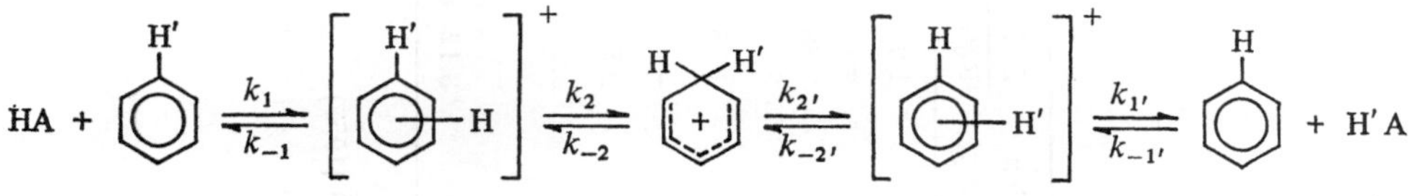

$$\text{(5)}$$

More recent experiments have given results other than the linear dependence of rate on H_0. Although the interpretation of these experiments is in dispute, there seems to be general agreement that the extent of proton transfer in the transition state and the residual interaction of the proton with A$^-$ is responsible for the results (*7*).

More information on the relative importance of complexes is provided by a comparison of the rate constants for substitution with the stability constants of the sigma and pi complexes (*19*).

Relative σ complex stabilities have been determined by measuring the distribution of methylbenzenes between the immiscible phases of a HF and heptane mixture (*20*). The solubility of the aromatic in the HF depends upon the extent of reaction (*6*):

$$\text{ArH} + \text{HF} \;\rightleftharpoons\; \text{ArH}_2^+ + \text{F}^- \qquad (6)$$

Relative π complex stabilities have been obtained by measuring the equilibrium (or argentation) constant of reaction (*7*):

$$\text{Ag}^+ + \text{ArH} \;\rightleftharpoons\; \text{AgArH}^+ \qquad (7)$$

The methylbenzenes are shaken for 12 hrs in equimolar $CH_3OH : H_2O$ which is 0.5 M in Ag$^+$ (*21*). These π stabilities are in substantial agreement with those obtained (*22*) by measuring the equilibrium of reaction (8):

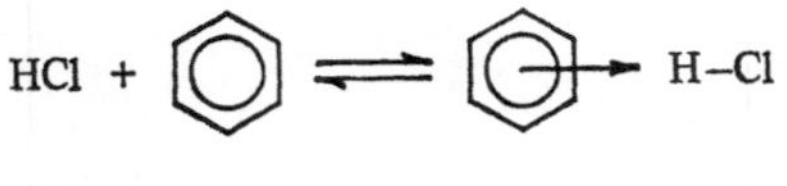

$$\text{(8)}$$

The difference between sigma and pi complex stabilities of various methylbenzenes, Table 1, is dramatic. The positive inductive effect of the methyl groups helps stabilize the positive charge developed in the σ complex, and accounts for the var-

Table 1. Comparison of relative δ and π complex stabilities with relative rates of substitution[a]

Benzene	δ complex stability	π Complex stability	Br_2 in 85% HOAc	Br_2 plus $FeCl_3$	Nitration $NO_2^+ BF_4^-$	Dedeuteration[b] Rate	Position of D
Unsubstituted	1.0	1.0	1.0	1.0	1.0	1.0	Equiv.
1-Methyl	7.9 × 10²	1.5	6.05 × 10²	3.6	1.6	3.1	2
1-Methyl						2.38 × 10²	3
1-Methyl						3.5 × 10²	4
1,4-Dimethyl	3.2 × 10³	1.6	2.5 × 10³	4.3	1.6	1.25 × 10³	Equiv.
1,2-Dimethyl	7.9 × 10³	1.8	5.3 × 10³	3.9	1.7	—	—
1,3-Dimethyl	1.0 × 10⁶	2.0	5.14 × 10⁵	5.6	1.9	2.5 × 10⁵	4
1,2,4-Trimethyl	2.0 × 10⁶	2.2	1.52 × 10⁶	—	—	—	—
1,2,3-Trimethyl	2.0 × 10⁶	2.4	1.67 × 10⁶	—	—	—	—
1,3,5-Trimethyl	6.3 × 10⁸	2.6	1.89 × 10⁸	—	2.7	3.12 × 10⁷[c]	Equiv.

a) Taken in part from Ref. (31), p. 502.
b) in H_2SO_4—CF_3CO_2H, Mackor, E. L., Smit, P. J., van der Waals, J. H.: Trans. Faraday Soc. 53, 1309 (1957).
c) in HOAc.

iation in the stabilities of the σ complexes. If the transition state resembles the σ complex, then a large increase in reaction rate should be observed upon introduction of methyl groups. This is the case for the reactions shown in columns 4 and 7 of Table 1. The π basicity of the benzene ring increases relatively little when methyl groups are added; thus, if the transition state resembles the π complex, little change in reaction rate should be observed. This is the case for the reactions shown in columns 5 and 6 of Table 1. According to this analysis it can be anticipated that dedeuteration of singly deuterated methylbenzenes by dilute H_2SO_4 in CF_3COOH (Table 1, Column 7) should involve a transition state that is stabilized by inductive electronic effects and is similar to a σ complex.

The rate constants for substitution of toluene compared to those of benzene, k_t/k_b, have also been used to observe such changes in the nature of the rate-determining transition state. By varying substituent X in $X-C_6H_4CH_2Cl$ and using a $TiCl_4-$ catalyzed benzylation, values for k_t/k_b from 2.5—97.0, were observed (23). These boundary values compare moderately well with values for the relative stability of π and of σ complexes of benzene and toluene shown in Table 1. They indicate a change in the character of the highest energy transition state from one resembling the σ complex (large k_t/k_b values) to one resembling the π complex or starting aromatics (low k_t/k_b values). This shift would seem to be the result of increased electrophilicity of the electrophile or increased reactivity of the aromatic, either of which lowers the energy of the transition state (and intermediates) relative to the reactants. The evidence from kinetic isotope effects; the correspondence between sigma—pi stabilities and relative reaction rates; and the toluene-benzene rate ratios, all point to a two-step reaction mechanism and the intermediacy of both sigma and pi complexes. Formation of the σ complex is required for hydrogen exchange to occur. When the rate-determining transition state is more similar to the σ complex, the formation of the π complex is viewed as a pre-reaction equilibrium. When the transition state is more similar to the π complex, then the σ complex is formed essentially irreversibly and goes over to product.

The mechanism of electrophilic substitution still remains clouded and caution must be exercised in extrapolation from one series of reactions to another. There is some recent evidence which even shows that the electrophile may, in some cases, attack the substituent rather than the ring.

C. Exchange Catalyzed by Protic Acids

In hydrogen exchange catalyzed by protic acids, in the absence of metal halides, formation of a σ complex-like transition state is generally believed to be the rate-determining step. Attempts have been made to correlate the differing rates of hydrogen exchange at various positions in polycyclic aromatic hydrocarbons with parameters based on quantum mechanical calculations (24—27). The most success-correlations have focused on cation localization energy, i.e., the amount of π-bonding energy required to isolate two electrons from the delocalized π system and locate them around one specific atom. The reaction rates for protiodetritiation (loss of ^3H) in CF_3CO_2H were measured for 21 positions in nine polycyclic aromatic hydrocarbons (28). Comparison of these rates with calculated values for cation

localization energy by CNDO/2 methods yielded a correlation coefficient of 0.979
(27). The correlation is thought to be due to the similarity between the theoretical
species 6, for which the calculations were made, and the transition state leading
to the σ complex 7.

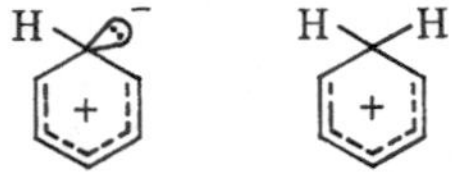

6   7

This relationship between structure and reaction rate permits, with some
exceptions (27), the prediction of relative positional exchange rates on previously
untested compounds.

D_2SO_4 is frequently used to replace a hydrogen by a deuterium atom in the
most reactive position(s) of a polycyclic compound. Thus treatment of the car-
cinogen, benzo[a]pyrene, 8, with concentrated D_2SO_4 for 2 minutes at 5—10°,
followed by quenching in D_2O and isolation of the product, showed that only
the 6-position, in accordance with calculations (26), had been deuteriodeprotonated:

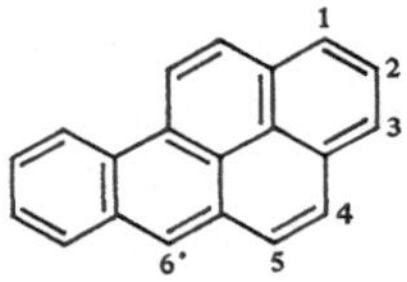

8

The position of substitution was demonstrated (29) by high resolution NMR.
When the 6-d compound was treated with H_2SO_4 under the above conditions,
protiodedeuteration occurred. The use of strong H_2SO_4 as the protic acid suffers
from the possibility of aromatic sulfonation. High positional selectivity depends on
very careful control of reaction conditions. Thus at room temperature, or on longer
treatment with D_2SO_4, other positions besides the 6-position of benzo[a]pyrene
become deuterated. In addition, some polycyclic hydrocarbons have limited
solubility in H_2SO_4. Trifluoracetic acid is preferred by some investigators for
incorporation of deuterium because of its high acidity and generally superior
solvent properties (25, 28). The acid-catalyzed hydrogen exchange reaction is the
reaction of choice for measuring the reactivities of the different positions of a
polynuclear hydrocarbon because of the absence of steric effects in both the
attacking electrophile and the leaving group.

D. Exchange Catalyzed by Lewis Acids

Addition of a Lewis acid to the reaction mixture can significantly increase the rate
of the hydrogen exchange process. [3]HCl will not exchange with the aromatic
hydrogens of toluene at temperatures up to 140°; upon addition of $SnCl_4$ the
reaction proceeds at room temperature (30). Lewis metal halides have been used to

catalyze hydrogen exchange reactions in both inorganic and organic acid systems.

The reason for the increased activity of protic acid-Lewis acid systems is ascribed to the enhancement of the protic acid acidity by the Lewis acid, (Table 2) (*31*). Whereas 100% HF has a H_0 value of —10.2, the acid system of HF +7% BF$_3$ has an H_0 value of —16.8, six orders of magnitude greater. The Hammett acidity function, H_0, is evaluated by the use of indicators, B, which are very weak bases and for which the acid ionization constants, Ka, of the conjugate acids, BH$^+$, have been determined (usually spectrophotometrically), Eq. (9):

$$BH^+ \rightleftharpoons B + H^+ \tag{9}$$

Table 2. H_0 values for HF, H$_2$SO$_4$, and some HF-Lewis acid mixtures

Acid	H_0
100% HF	— 10.2
100% H$_2$SO$_4$	— 10.6
HF + 0.36M NbF$_5$ (6.7% weight)	— 13.5
HF + 3M SbF$_5$ (60% weight)	— 15.2
HF + 1M BF$_3$ (7% weight)	— 16.8

H_0 is defined by Eq. (10):

$$H_0 = \mathrm{pK}_a - \log \left[\frac{BH^+}{B} \right] \tag{10}$$

where $\left[\dfrac{BH^+}{B} \right]$ is the experimentally observed ratio of the concentration of the indicator in its two forms. By establishing a series of sequentially weaker bases it is possible to measure the proton-donating ability of very strong acidic media by this technique. The H_0 scale as defined above is consistent with the pH scale, in that the stronger the acid, the smaller its H_0 value.

It has been traditional to refer to water or hydrogen halides as co-catalysts to the Friedel-Crafts catalyst. The function of the Lewis acid is to react with the HX and thereby enhance the acidity of the protic acid. For this reason, the Lewis acid can be viewed as a co-acid. Studies with rigorously purified materials have shown that in the absence of water or a hydrogen halide, Lewis acids do not catalyze hydrogen exchange (*32*) and that a conducting species (the σ complex) is not formed; only weak π interaction can be observed (*33*). Addition of AlBr$_3$ to a solution of mesitylene in liquid DBr results in a 10^4 increase in specific conductivity and initiates hydrogen exchange. In the absence of the aromatic hydrocarbon, AlBr$_3$ in liquid HBr is a very poor conductor (*32*). There is some evidence that the proton-donating species, which is present in the mixtures of protic acids with Lewis acids used for hydrogen transfer to the aromatic, is formed by the general reaction (11):

$$2 \, HX + MX_n \rightleftharpoons H_2X^+ + MX_{n\pm1} \tag{11}$$

This equilibrium has been observed in conductivity studies for reaction (12):

$$2 \, HF + SbF_5 \rightleftharpoons H_2F^+ + SbF_6^- \tag{12}$$

Three separate investigators have determined that the molar ratio of 2:1 in the complex acid system $CH_3COOH:SnCl_4$ is the most active ratio for catalyzing hydrogen exchange in aromatic hydrocarbons (34). BF_3 has also been observed to form a 1:2 molar complex with acetic acid (35).

Using a variety of metal bromides as Lewis catalysts, rate constants have been determined for the hydrogen exchange of benzene (32) in DBr. The activity order is $AlBr_3 > GaBr_3 > FeBr_3 > BBr_3 > SbBr_3 > TiBr_4$ with a reaction rate range of 10^5 from the fastest to the slowest. This activity order parallels the previously determined acidity of the corresponding metal chlorides (36), and indicates the importance of acidity in the catalytic role of Lewis acids. The rates of H—D exchange for benzoic acid and nitrobenzene catalyzed by $AlBr_3$ were found to be 10^{-4} and 10^{-6}, respectively, of the rate with benzene (32). Such results are consistent with the known effect of electron-withdrawing substituents and suggest that the transition states for the reaction have σ complex character.

Hydrogen exchange of aromatics, including exchange of all aromatic hydrogen positions in larger polycyclic aromatics, can be achieved by cross-exchange with benzene-d_6 in the presence of a Lewis acid (37, 38). Virtually complete equilibration between all aromatic and all acidic hydrogens in the system is observed at room temperature for one and two-ring aromatics. No exchange in alkyl side chains is observed. Most likely, the Lewis acid reacts with the trace quantities of water in the benzene-d_6 to form hydrogen halide. In the case of $AlBr_3$ the reaction proceeds according to Eq. (13):

$$AlBr_3 + nH_2O \rightleftharpoons nHBr + AlBr_{3-n}(OH)_n \tag{13}$$

The liberated HBr reacts with $AlBr_3$ (probably as the dimer) in the presence of benzene to form the sigma complex, Eq. (14):

$$HBr + AlBr_3 + C_6D_6 \longrightarrow C_6D_6H^+ \, AlBr_4^- \tag{14}$$

Reversal of reaction (14) yields DBr in solution which cycles through the same reaction with the substrate aromatic compound to yield the deuterated product. In a typical experiment, conducted in the authors' laboratory, reaction of 0.050 g of pyrene with 1 ml benzene-d_6 and 0.040 g $AlBr_3$ in a sealed tube for 5 days at 90° gave pyrene in which 92 percent of all hydrogens had been exchanged for deuterium (mass spectrometry). A second treatment yields about 99 percent exchange of all hydrogens.

179

Ethylaluminium dichloride is an equally effective catalyst in cross-exchanging deuterium from donor C_6D_6; however, at the higher temperatures required to achieve exchange in larger polycyclics, some ethylation of the aromatic was observed (work performed in the author's laboratory).

III. Base-Catalyzed Hydrogen Exchange

A. Benzyne Intermediates

The early chemistry dealing with the reaction of aromatic halides with strong bases is replete with examples of rearrangement; *i.e.*, the entering group frequently does not take up the position of the displaced halogen. Thus, for example, treatment of *o*-chloroanisole *9* with $NaNH_2$ in liquid NH_3 (—33°) at atmospheric pressure gives the *m*-anisole *10*, exclusively (15):

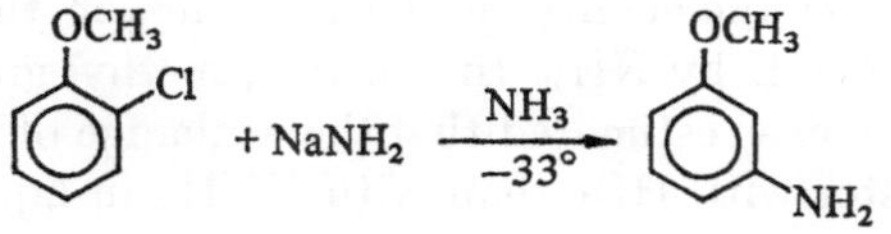

$$(15)$$

It is known now that this and similar reactions proceed by a two-step mechanism involving first; the removal of a proton by the base, followed by amination of the benzyne intermediate, *11*, (16):

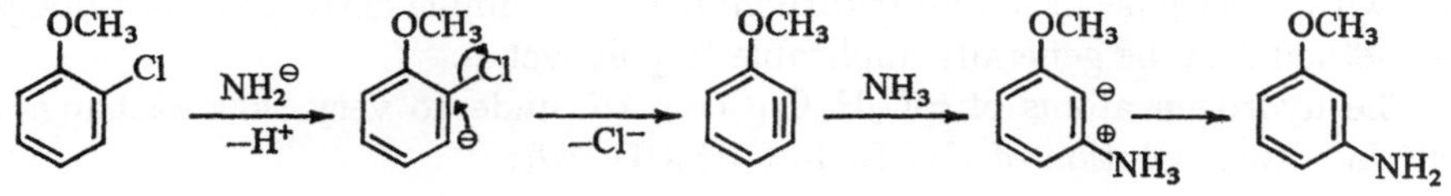

$$(16)$$

That this was indeed the mechanism was shown conclusively by treating chlorobenzene, labelled with ^{14}C at the position of substitution, *12*, with KNH_2 in liquid NH_3. The product was 50 percent aniline labelled at the amino group, *13*, and 50 percent aniline labelled at the position ortho to the amino group, *14*, (17):

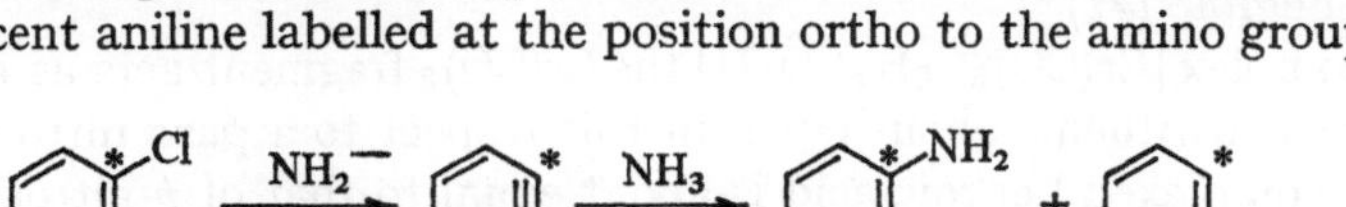

$$(17)$$

It is apparent that if the strong base attacks the aromatic compound in an acid-base equilibrium reaction which is fast, compared to subsequent reactions, the halobenzene will undergo hydrogen exchange. Thus fluorobenzene-2,6-d_2, *15*, exchanges with hydrogens when treated with KNH_2 in liquid NH_3:

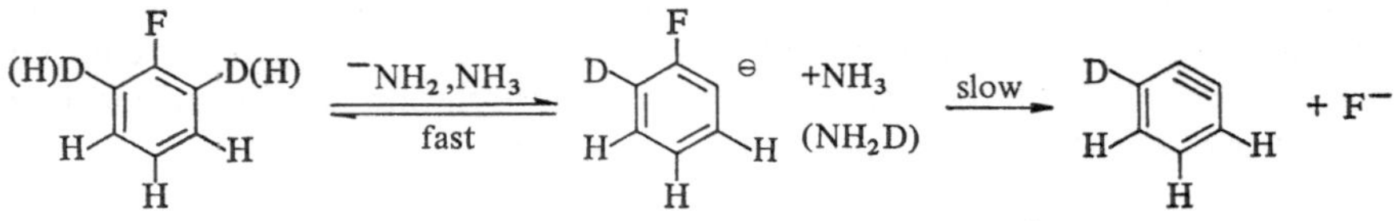

15

(18)

When C_6D_6 is treated with NH_2^- in NH_3 it is possible to exchange all D with H.

Some interesting work has been done on the effect of substituents on the rate of deuterium-hydrogen exchange in simple benzene derivatives. Thus it has been shown (39) that increasing the number of methyl groups on deuteriobenzene lowers the rate at which the deuterium exchanges with H in the presence of 0.2N KNH_2 in liquid NH_3 at 25°. This is expected since the methyl group is an electron-donating group and hence reduces the acidity of the D atoms on the ring, making them more difficult to remove D by NH_2^-. In contrast, methyl groups enhance electrophilic exchange. It has been estimated that the exchange of D by H in $C_6D(CH_3)_5$ would occur 10^{12} faster with HBr than with KNH_2 in liquid NH_3 (39).

B. Arenechromium Tricarbonyls

Although π-benzenechromium tricarbonyl, π-$C_6H_6Cr(CO)_3$, is one of the best characterized arenechromium tricarbonyls, other arenes also form complexes with $Cr(CO)_3$. Accordingly, although isotopic exchange apparently has been investigated only with π-benzenechromium tricarbonyl, and its simple derivatives in principle the method may be generally applicable to polycyclics.

The hydrogen atoms of π-$C_6H_6Cr(CO)_3$, 16, undergo very slow exchange at 100° in a 10% solution of NaOEt in C_2H_5OD (40):

$$\pi\text{-}C_6H_6Cr(CO)_3 \underset{}{\overset{\text{EtOD, EtONa}}{\rightleftharpoons}} \pi\text{-}C_6D_6Cr(CO)_3$$

16

(19)

The effect of substituents on the benzene ring on the rate of isotopic exchange is rather negligible (41).

In the complex [$Cr(CO)_3C_6H_5CO_2H$] the $Cr(CO)_3$ fragment acts as an electron-withdrawing substituent, about equal in this respect to a *para* nitro group. The pKa of the complexed benzoic acid is about equal to that of p-nitrobenzoic acid (42). Consistent with the suggested electron-withdrawing effect of $Cr(CO)_3$ complexed to an aromatic, it was found that a complexed halobenzene undergoes nucleophilic substitution with methoxide faster than dole free halobenzene (43).

As might be expected from the results of base-catalyzed exchange of π-$C_6H_5Cr(CO)_3$, certain cyclopentadienyl complexes also undergo exchange. All the hydrogens of the cyclopentadiene ring of π-$C_5H_5Mn(CO)_3$ and π-$C_5H_5Re(CO)_3$ exchange with deuterium on treatment with C_2H_5OD in the presence of C_2H_5ONa. Again the reaction is slow and must be carried out in a closed vessel at 100° (44).

M. Orchin and D. M. Bollinger

C. Benzylic Chromium Tricarbonyls

The benzylic protons of π-alkylarene $Cr(CO)_3$ complexes can be completely exchanged by treating the complexes with $(CH_3)_3COK$ in dimethylsulfoxide-d_6
(45). The benzylic protons in uncomplexed alkylbenzenes are known to be acidic;
consistent with the electron-withdrawing effect mentioned above, complexation
with $Cr(CO)_3$ enhances this acidity. The reaction sequence probably consists of
the following steps (20):

$$
\begin{array}{l}
\underset{|}{Cr(CO)_3} \\
ArCH_2R + (CH_3)_3CO^- \;\rightleftharpoons\; \underset{|}{\overset{Cr(CO)_3}{Ar\bar{C}HR}} + (CH_3)_3COH \\[2mm]
\underset{|}{\overset{Cr(CO)_3}{ArC\bar{H}R}} + CD_3SOCD_3 \;\rightleftharpoons\; \underset{|}{\overset{Cr(CO)_3}{ArCHDR}} + \bar{C}D_2SOCD_3 \\[2mm]
(CH_3)_3COH + \bar{C}D_2SOCD_3 \;\rightleftharpoons\; (CH_3)_3CO^- + CD_2HSOCD_3
\end{array}
\tag{20}
$$

The similar exchange reaction with π-indanechromium tricarbonyl, 17, led to
the exchange of only the two *anti* hydrogen atoms. 18, indicating the great stereo-
selectivity of this reaction:

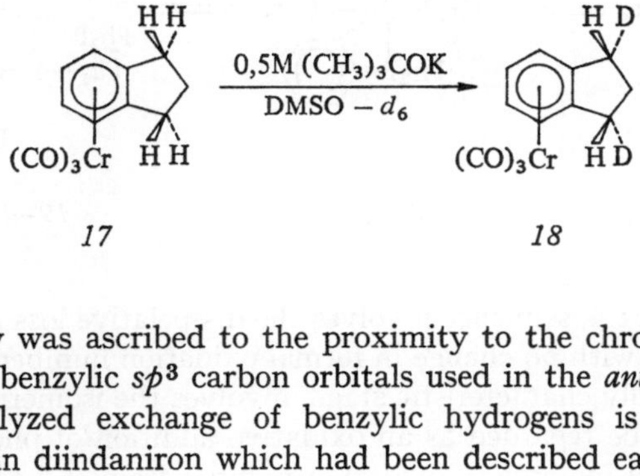

$$
\tag{21}
$$

The selectivity was ascribed to the proximity to the chromium atom of the
back lobes of the benzylic sp^3 carbon orbitals used in the *anti* C—H bonds.

The base-catalyzed exchange of benzylic hydrogens is analogous to the
similar exchange in diindaniron which had been described earlier (46).

IV. Exchange of Ortho Hydrogens in Aromatic Phosphine Ligands of Metal Complexes

The development of the new field of organometal chemistry during the last two
decades has led to the discovery of many unusual organometal complexes which
are soluble in organic solvents. The ability of some of these complexes to react
with molecular oxygen or with molecular nitrogen under very mild conditions has
been of particular interest because the complexation of these simple but ubiquitous molecules to transition metals constitutes the first step in two of the most

182

fundamental reactions which characterize life on this planet, namely, the oxygenation of blood and the fixation of nitrogen. The complexation of another simple molecule, molecular hydrogen, to soluble transition metal complexes mimics the first step in another fundamental reaction of considerable interest to the chemist — the heterogeneous catalyzed hydrogenation of unsaturated compounds. Transition metal complexes which have aromatic phosphines coordinated to them are of particular interest in this connection. Some of these complexes are capable of complexing and activating (splitting) molecular hydrogen and accordingly provide an unusual example of isotopic exchange.

The essential series of reactions leading to hydrogen-deuterium exchange in aromatic phosphines can be illustrated with the chlorohydrido-*tris*-triphenylphosphineruthenium(II) complex, *19*:

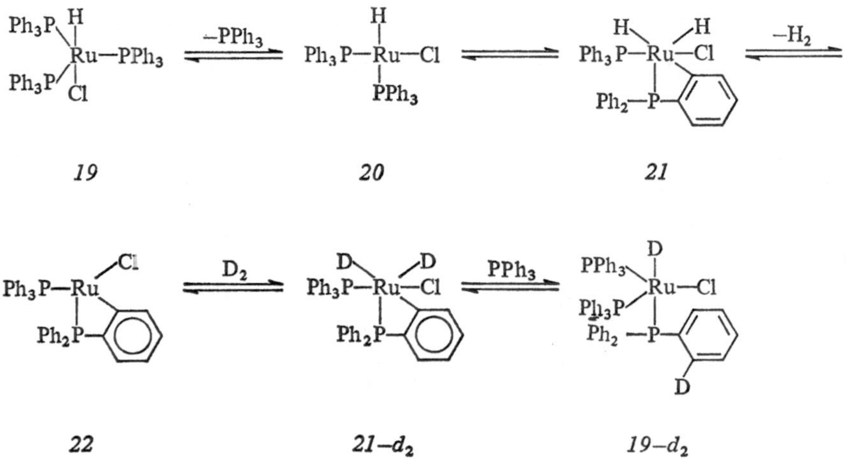

$$(22)$$

The first step in the sequence involves the dissociative loss of a neutral ligand from *19* to form *20* with no change in formal oxidation number of the metal. The second, and especially characteristic stage, involves the isomerization of *20* to *21*. This reaction may be regarded as an oxidative addition of phenyl and H to the metal with a formal increase in oxidation number from $+2$ to $+4$; it may also be viewed as metal insertion into a C—H sigma bond. Such an isomerization is analogous to an earlier established example (*47*) of this type of reaction:

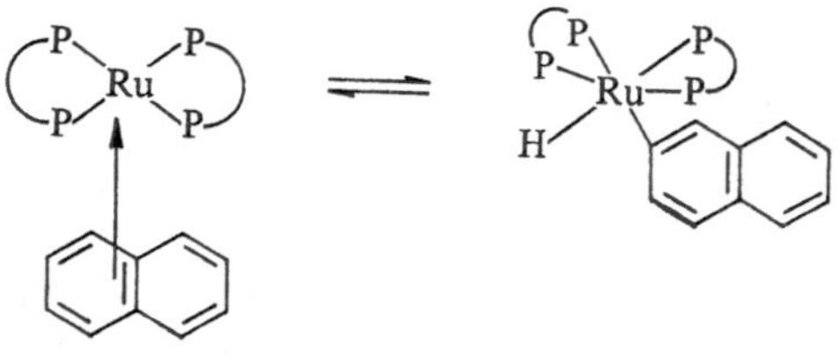

$$(23)$$

$$\widehat{P\,P} = (CH_3)_2PCH_2CH_2P(CH_3)_2$$

The conversion of *21* to *22* with loss of H_2 may be viewed as a reductive elimination and the equilibrium between *21* and *22* provides the reaction which leads to H_2—D_2 exchange. The reductive elimination reaction *21*-d_2 → *19*-d_2 results in the incorporation of one ortho D atom. Because *19* has 18 such ortho protons in addition to the Ru—H proton, it is possible to exchange 19 H atoms in *19* with D atoms in the presence of a large excess of D_2 and indeed $[(C_6H_5\text{-}2,6\text{-}d_2)_3P]_3$ RuDCl has been isolated (*48*).

[RuHCl(PPh$_3$)$_3$], *19*, is an excellent hydrogenation catalyst for terminal olefins because, among other reasons, it readily loses a ligand PPh$_3$ in solution, thus opening a site for the olefin. The exchange of the ortho hydrogens with D is rapid even at room temperature and it is known that the coordinated phosphines are in rapid equilibrium with free ligand in solution. Indeed if the exchange with D_2 is carried out in the presence of free $(C_6H_5)_3P$, $(C_6H_5\text{-}2,6\text{-}d_2)_3P$ can be isolated (*48*). Exchanges similar to that shown above for *19* also occur with the dinitrogen complexes [CoH(N$_2$)(PPh$_3$)$_3$] and [RuH$_2$(N$_2$)(PPh$_3$)$_3$] as well as with some other similar complexes.

It is interesting that the complex [IrCl(PPh$_3$)$_3$], *23*, although it readily and irreversibly forms a dihydride adduct with H_2, is not a good hydrogenation catalyst like [RhCl(PPh$_3$)$_3$] or [RuHCl(PPh$_3$)$_3$], presumably because the Ir complex does not readily lose a ligand in solution. However, on heating *23* in benzene solution it isomerizes with oxidative addition to the hydridochloro species:

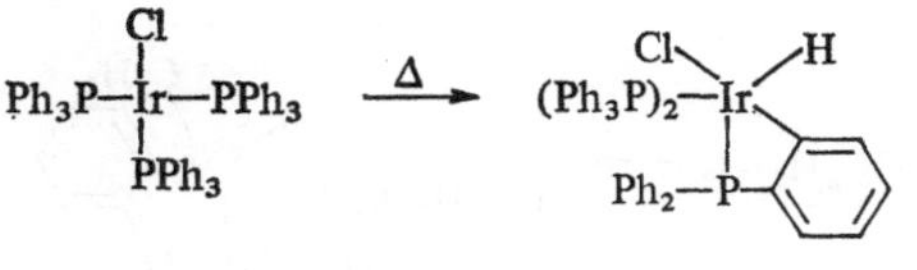

$$\begin{array}{c}
\underset{\displaystyle *23*}{\underset{\displaystyle \overset{\displaystyle Cl}{|}}{Ph_3P-\overset{\displaystyle |}{Ir}-PPh_3}} \xrightarrow{\Delta} (Ph_3P)_2-Ir \cdots
\end{array} \qquad (24)$$

This is not a particularly good system for obtaining H—D exchange of all the ortho hydrogens on the phosphines (*49*).

In the reductive elimination steps we have discussed above, *e.g.*, *21* → *22*, either H_2, HD, or D_2 is eliminated. If in addition to a hydrogen atom, there is an alkyl group attached to the metal by a sigma bond, it is possible to lose RH. Thus it has been shown (*50*) that [Rh(CH$_3$)(PPh$_3$)$_3$], *24* readily loses CH_4. The loss of methane probably proceeds as follows:

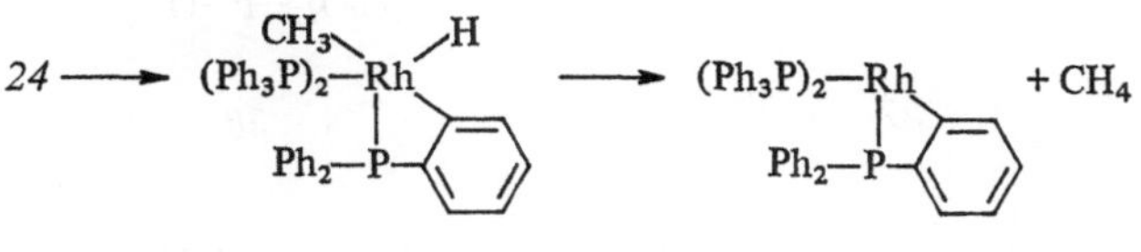

$$*24* \longrightarrow \underset{\displaystyle *25*}{(Ph_3P)_2-Rh} \longrightarrow \underset{\displaystyle *26*}{(Ph_3P)_2-Rh} + CH_4 \qquad (25)$$

The reductive elimination step also probably occurs during the final step in catalytic hydrogenations with active organometal catalyst. Thus the hydrogenation of ethylene with *19* may be written as occurring through the following sequence:

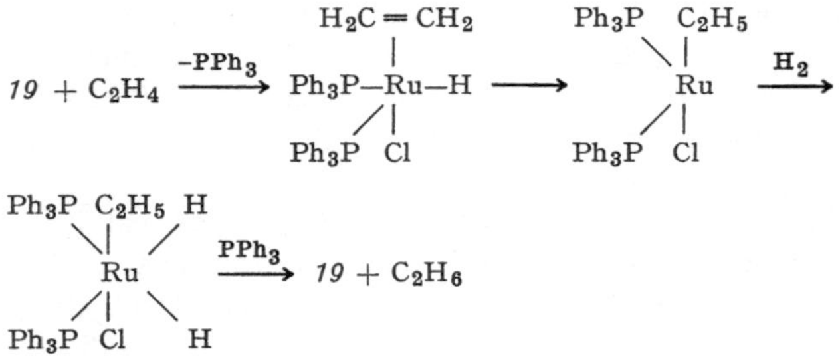

$$19 + C_2H_4 \xrightarrow{-PPh_3} \underset{\substack{| \\ Ph_3P \quad Cl}}{Ph_3P-Ru-H} \longrightarrow \underset{\substack{/ | \\ Ph_3P \quad Cl}}{\overset{\substack{Ph_3P \quad C_2H_5 \\ \backslash |}}{Ru}} \xrightarrow{H_2}$$

$$\underset{\substack{/ | \quad \backslash \\ Ph_3P \quad Cl \quad H}}{\overset{\substack{Ph_3P \quad C_2H_5 \quad H \\ \backslash | \quad /}}{Ru}} \xrightarrow{PPh_3} 19 + C_2H_6 \tag{26}$$

The stoichiometric hydrogenation of olefins with a catalyst such as [NiBr$_2$(PPh$_3$)$_2$] (*51*), in the absence of added hydrogen can be rationalized by utilization of the ortho hydrogens as the hydrogen source (*52*). This would be followed by reductive elimination of alkane, analogous to the conversion of *25* → *26*, illustrated above. Additional evidence for ligand-metal hydrogen exchange comes from studies (*53*) of H—D exchange in C$_2$D$_4$ catalyzed by CoH[P(OC$_6$H$_5$)$_3$]$_4$. The quantity of hydrogen-containing ethylenes was higher than the theoretical value calculated for exchange of only the Co—H. In addition some ethane was formed. Both observations can be reconciled with utilization of ortho hydrogens of triphenyl phosphite:

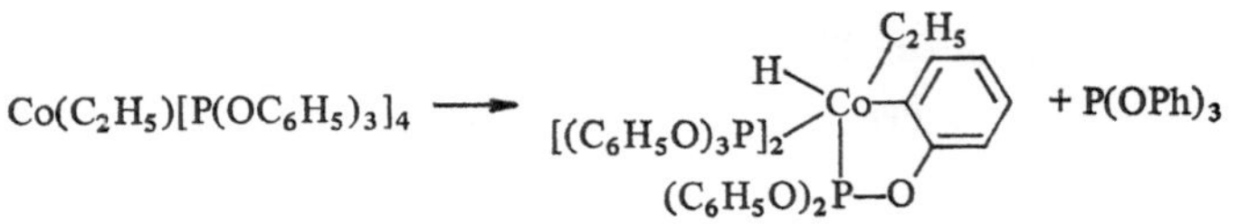

$$Co(C_2H_5)[P(OC_6H_5)_3]_4 \longrightarrow [(C_6H_5O)_3P]_2 \underset{(C_6H_5O)_2P-O}{\overset{H\diagdown \quad \diagup C_2H_5}{Co}} + P(OPh)_3 \tag{27}$$

The transfer of ortho hydrogen from phenylphosphine to the coordinated metal atom (ligand-metal hydrogen transfer) involves a 4-membered ring intermediate, *e.g.*, *22* or *26*. Coordinated triphenylphosphites also readily undergo ortho hydrogen exchange and here a 5-membered ring is involved. Thus there is a facile equilibrium between *27* and *28*.

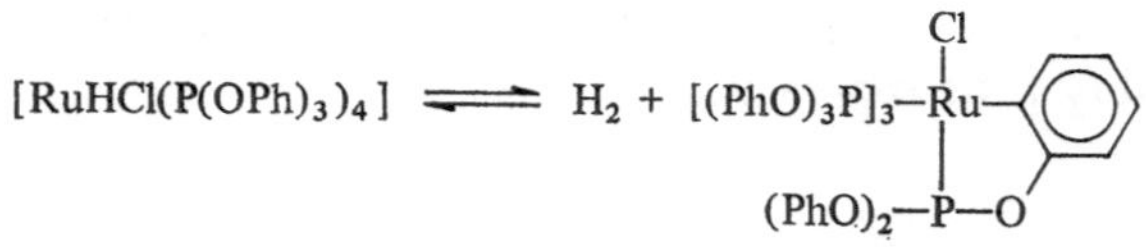

$$[RuHCl(P(OPh)_3)_4] \rightleftharpoons H_2 + \underset{(PhO)_2-P-O}{\overset{Cl}{[(PhO)_3P]_3-Ru}} \tag{28}$$

$$\quad\quad 27 \quad\quad\quad\quad\quad\quad\quad\quad\quad\quad 28$$

In the presence of deuterium, the 24 ortho hydrogens of *27* are exchanged for D at room temperature (*48*). It is pertinent to note that *27* undergoes a reductive elimination of H$_2$ rather than HCl. When phenol was added to solutions of *27*, the phenol was also ortho deuterated indicating that phenol was exchanging with the phenoxy groups of the triphenylphosphite.

The substitution of ortho hydrogens of coordinated phenylphosphines and phenylphosphites suggests that the formation of the cyclic intermediate proceeds via a Friedel-Crafts-type electrophilic substitution (52). The complete scheme, using 27 ⇌ 28 as an example, may be written as follows:

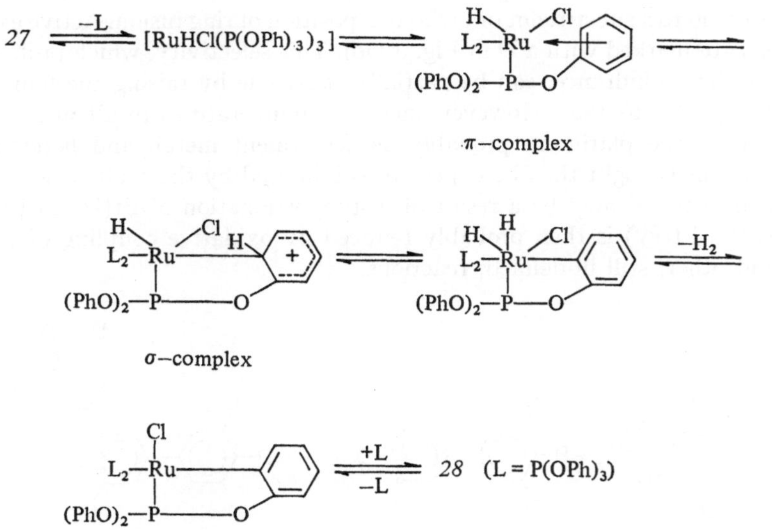

$$27 \; \xrightleftharpoons{-L} \; [RuHCl(P(OPh)_3)_3] \rightleftharpoons \quad \underset{(PhO)_2-P-O}{\overset{H\quad Cl}{L_2-Ru}} \rightleftharpoons$$

π−complex

$$\underset{(PhO)_2-P\text{———}O}{\overset{H\quad Cl\quad H}{L_2-Ru}} (+) \rightleftharpoons \underset{(PhO)_2-P-O}{\overset{H\; H\; Cl}{L_2-Ru}} \xrightarrow{-H_2} \qquad (29)$$

σ−complex

$$\underset{(PhO)_2-P\text{———}O}{\overset{Cl}{L_2-Ru}} \xrightleftharpoons[-L]{+L} \; 28 \quad (L = P(OPh)_3)$$

The dissociative loss of a ligand in the first step allows the arene to π complex with the metal. In support of the electrophilic substitution mechanism it has been shown (49) that electron-donating substituents on the ring enhance the rate of ortho substitution. In the above example, this would be consistent with the attack of the electrophile Ru^{2+} in the π complex to form the σ complex which is then stabilized by proton transfer (oxidation) to form the dihydrido species. The essential steps in the reaction sequence are all reasonable and each is more or less documented and hence there is good reason to believe that H—D exchange occurs essentially in the manner shown.

Many other phosphine complexes, in addition to those mentioned in this short account, have been reported to undergo exchange. For a more complete account see Ref. (52).

V. Exchange Catalyzed by Soluble Platinum Catalysts

A. General

Solutions of platinum(II) chlorides in $D_2O:CH_3CO_2D$ have been shown (54—56) to be effective catalysts for exchanging the hydrogens of benzene, alkylbenzenes, polyphenyls, and polycyclic hydrocarbons. In a typical experiment with biphenyl, 0.11 g (0.7×10^{-3} mol) of the hydrocarbon, 0.50 g (1.30×10^{-3} mol) of Na_2PtCl_4 and 0.048 g (1.30×10^{-3} mol) of DCl (generated by reacting acetyl chloride with D_2O) are added to 2 ml of a D_2O solution containing 67 mole percent CH_3CO_2D.

The solution is heated at 100° for 5 h (*56*). Under these conditions, Na₂PtCl₄ catalyzes the exchange of 41.2 percent of the hydrogens in the biphenyl and under similar conditions, 36.1 percent of the hydrogens in a 0.105 g (0.82×10^{-3} mol) sample of naphthalene.

Exchange proceeds at all positions in aromatic compounds with the exception of those ortho to a substituent or ortho to a position of ring fusion. Active exchange positions are marked with a D in Fig. 3 (*56*). This selectivity, which probably is a result of steric hindrance, can be partially overcome by raising the temperature from about 100° to 130°. However, increased temperatures result in a slow precipitation of the platinum, probably as zero-valent metal, and heterogeneous exchange which might then be expected, is inhibited by the acetic acid. The precipitation of metal may be a result of disproportionation of Pt(II) to Pt(0) and Pt(IV); the Pt(IV) is then probably reduced by oxidative coupling of aromatic rings and other, still ill-defined, reactions.

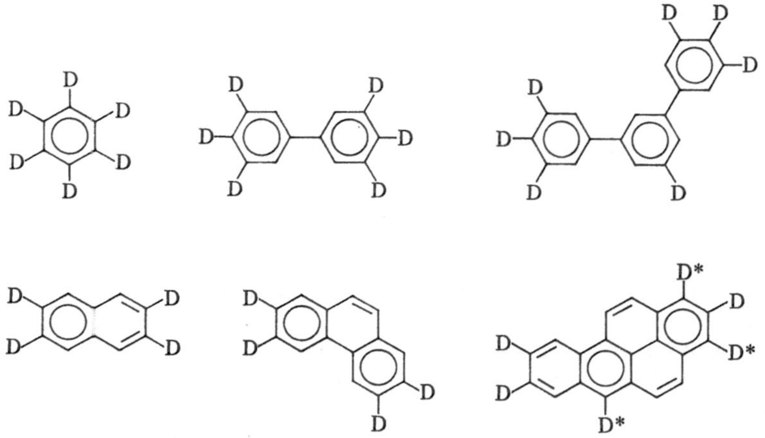

Fig. 3. Active positions for H—D exchange. [* Substitution at these positions is most likely acid catalyzed (*29*)]

B. Mechanism

A suitable mechanism for the homogeneous platinum catalyzed exchange must rationalize two important experimental observations: (1) low reactivity at the ortho positions of substituted benzenes and (2) absence of substantial electronic effects of substituents in the benzene ring (*57*). For these reasons, a mechanism involving the direct attack of external D⁺ on the "π complexed" benzene has been rejected.

Two independent pathways for the exchange mechanism, both of which accommodate experimental observations, have been proposed (*57*) and these are shown in Fig. 4. Pathway I may be called the oxidative addition pathway. It involves attack on *29* to give a 5-coordinate Pt(II) π-complex intermediate (or transition state), *30*. Complex *30* then undergoes the oxidative addition (or metal insertion into an aromatic C—H bond) to give the Pt(IV) complex, *31*. This step is exactly analogous to the Ru—$C_{10}H_8$ case mentioned earlier [Eq. (23)]. Reductive elimination converts *31* to the σ-bonded complex *34* with extrusion of HCl. The oxidative addition of DCl and its reductive elimination giving the equilibria between *34*, *31*, and *30* provides for the H—D exchange. The deuterated aromatic is eventually released from *30* by its displacement from the catalyst in the equilibrium reaction *30* ⇌ *29*.

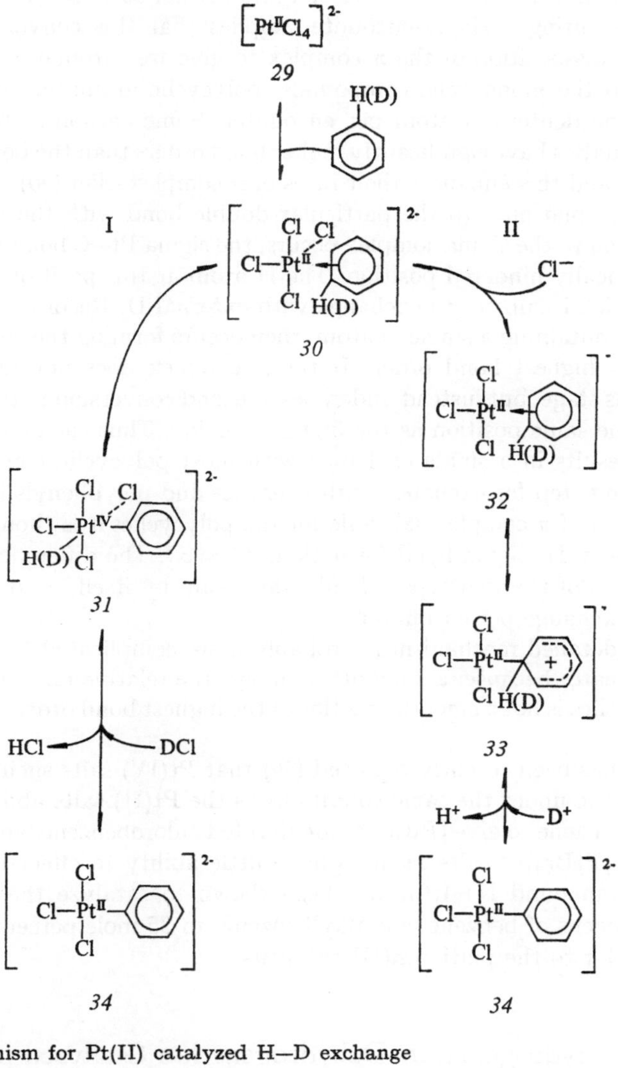

Fig. 4. Mechanism for Pt(II) catalyzed H—D exchange

Pathway II involves the conversion of *30* to the π-complex intermediate, *32*. The conversion of *32* to *33* may be considered as an electrophilic attack by Pt(II) on the aromatic nucleus to give the usual delocalized cationic intermediate characteristic of aromatic electrophilic substitution reactions. The loss of a proton from *33* to give *34* (and the reverse addition) provides the step for H—D exchange. At the present time it appears that a choice between pathways must await further experiments although some preference has been expressed for Pathway I (*58*).

Benzene, alkylbenzenes, and polyphenyls appear to undergo multiple exchange, *i.e.*, the introduction of two or three deuterium atoms per encounter with the platinum atom. It will be noted that both pathways I and II require that each single exchange involves a π to σ conversion, *i.e.*, either *30* to *34* or *32* to *34*. Multiple exchange during a single encounter implies that this conversion is rapid relative to the dissociation of the π complex to give free aromatic.

Contrary to the monocyclic compounds, polycyclic aromatics appear to exchange only one deuterium atom per encounter. Some carbon-carbon bonds in polycyclic aromatics have significantly higher bond orders than the bonds in single-ring aromatics and this enhances their rates of π complexation (*56*). After π complexation of the platinum to the particular double bond with the highest bond order, conversion to the sigma complex occurs, the sigma Pt—C bond being formed at the less sterically hindered position. The H atom at this position becomes the labile atom which is subject to exchange with external D. Reconversion to the π complex, now containing a single D atom, then occurs forming the complex at the original site of highest bond order. If the π complex does not release the d_1 aromatic at this stage, but instead undergoes a second conversion to the σ complex, it does so at the same position as the first conversion. Thus each encounter with Pt generally results in a single exchange with most polycyclic compounds. The rate determining step for exchange with benzenes and polyphenyls is thought to be the formation of π complex *32*, while for the polycyclics it is thought to be, at least via pathway II, sigma bond formation, *32* ⇌ *33*. The difference in the rate determining step of the two types of substrates can, by itself, account for single and multiple exchange per encounter.

The exact detailed mechanism is probably more complicated than the above discussion indicates because, among other things, the relative rate of exchange of various polycyclics is not a smooth function of the highest bond order of the various polycyclics (*56*).

Finally, it has been recently reported (*59*) that Pt(IV) salts such as $[PtCl_6]^{2-}$ deuterate benzene under the same conditions as the Pt(II) salts above. $[PtCl_6]^{2-}$ is reduced by benzene to give $[PtCl_4]^{2-}$, deuterated chlorobenzene, and deuterated biphenyl. The $[PtBr_4]^{2-}$ salts seem to have little ability to effect hydrogen exchange. The compound Na_3IrCl_6 has been shown to catalyze the exchange of aromatic hydrogens of benzene and alkylbenzenes in 25 mole percent CH_3CO_2D: D_2O, (*60*) similar to the platinum(II) catalysts.

VI. Selective Hydrogenation–Dehydrogenation Catalyzed by Dicobalt Octacarbonyl

Although ordinarily the system $Co_2(CO)_8$ plus high pressure synthesis gas ($H_2 +$ CO) results in hydroformylation of olefinic bonds, *i.e.*, production of aldehydes, certain substrates that do not possess isolated double bond character can be hydrogenated rather than hydroformylated (67). Thus when anthracene is treated with H_2 and CO at high pressures in the presence of catalytic quantities of $Co_2(CO)_8$, quantitative yields of 9,10-dihydroanthracene are formed (62). It is known that the active catalyst under these catalytic oxo or hydroformylation conditions is $HCo(CO)_4$ and indeed when anthracene, 35, is treated with this carbonyl at room conditions, 9,10-dihydroanthracene, 36, is formed (63).

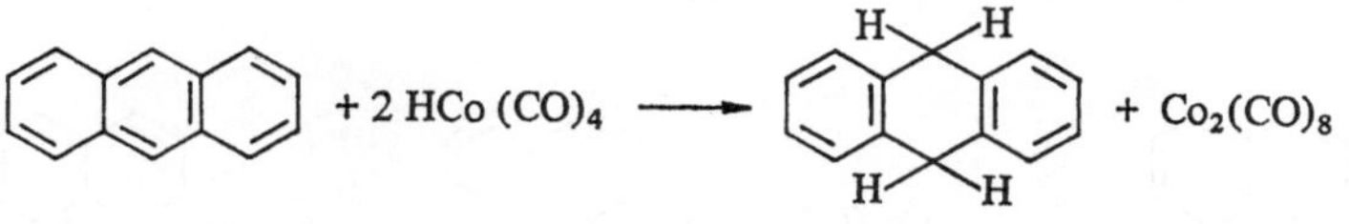

$$\text{(structure 35)} + 2\ HCo(CO)_4 \longrightarrow \text{(structure 36)} + Co_2(CO)_8 \qquad (30)$$

35   36

If a benzene solution of anthracene is treated with CO (1350 psi) $+D_2$ (1300 psi) in the presence of $Co_2(CO)_8$, at 200° for 11 hrs, the tetradeuterio compound 36-d_4 is formed in good yield (64). The tetradeuterio compound, 9,10-dihydro-anthracene-9,9,10,10-d_4, probably arises from a series of hydrogenation—dehydrogenation reactions which, neglecting the equilibria involving H—D exchange, may be written:

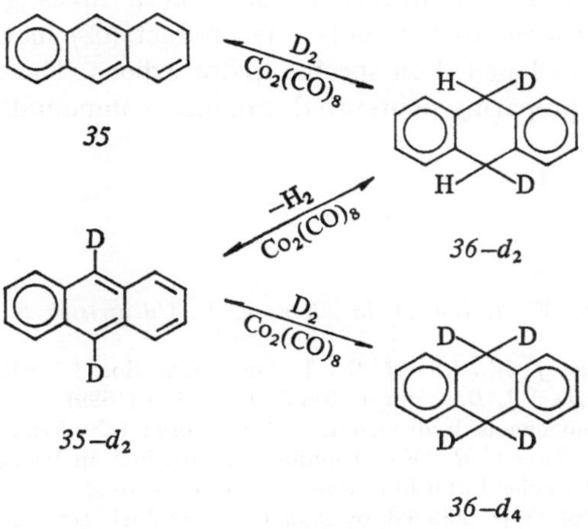

$$\text{(reaction scheme: 35 / 35-}d_2 \rightarrow 36\text{-}d_2 / 36\text{-}d_4\text{)} \qquad (31)$$

35

36-d_2

35-d_2

36-d_4

Under the above conditions, the d_4 compound is formed in 71 percent yield. The other major product is the monoprotio analog of 36 (22%). Careful characterization of the products showed no D on the outer aromatic rings. Although such a conversion has not been reported, if 36-d_4 were to be dehydrogenated chemically, with a reagent such as sulfur or chloranil, it would be possible to obtain 35-d_2.

In a convenient modification of the above procedure (64), it is possible to use D_2O-dioxane as a source of deuterium. Thus if anthracene is dissolved in D_2O-dioxane and treated with high pressure (3000 psi) CO at 175°, 36-d_4 can be formed. The D_2O probably reacts with CO in the water-gas shift reaction:

$$CO + D_2O \overset{Co_2(CO)_8}{\rightleftharpoons} D_2 + CO_2 \qquad (32)$$

$$D_2 + Co_2(CO)_8 \rightleftharpoons 2\ DCo(CO)_4 \qquad (33)$$

Since D_2O is considerably cheaper and more convenient to use than D_2, this is an attractive method to place D specifically at the 9,10 positions of anthracene. Pyrene, 37, is also selectively hydrogenated by the $Co_2(CO)_8 + CO + D_2$ system, providing a possible route to the specifically labelled pyrene-4,5-d_2, 37-d_2. Retreatment of 37-d_2 might lead to pyrene-4,5,9,10-d_4.

$$(34)$$

37   37–d_2

The above reaction incorporating D is obviously restricted to those polycyclic hydrocarbons that react in the $Co_2(CO)_8$ catalytic system; but not all polycyclics are reactive; e.g., phenanthrene reacts only very slowly under these conditions (65). Quite a few hydrocarbons have been investigated. When they react they generally seem to give only one product (62) making it possible to have specific positions labelled on specific hydrocarbons. This is obviously not a general method for preparing deuterated aromatic compounds.

VII. References

1. *Guroff, G., Daly, J. W., Jerina, D. M., Renson, J., Udenfriend, S., Witkof, B.*: Science 157, 1524 (1967).
2. *Jerina, D. M., Daly, J. W., Witkof, B.*: J. Am. Chem. Soc. 90, 6523 (1968).
3. *Sabatier, P., Senderens, J. B.*: Compt. Rend. 128, 1173 (1899).
4. *James, B. R.*: Homogeneous hydrogenation. New York: John Wiley and Sons 1973.
5. *Norman, R. O. C., Taylor, R.*: Electrophilic substitutions in benzenoid compounds, p. 203. Amsterdam: Elsevier Publishing Company 1965. — *Gold, V.*: In: Friedel-Crafts and related reactions, Vol. II, p. 1259 (ed. by *Olah, G.*). New York: Interscience Publishers 1964.
6. *Wiberg, K. B.*: Chem. Rev. 55, 713 (1955).
7. *Berliner, E.*: In: Prog. Phys. Org. Chem., Vol. 2, p. 253 (ed. by *Cohen, S. G., Streitweiser, Jr., A., Taft, R. W.*). New York: Interscience Publishers 1964.
8. *Westheimer, F. H.*: Chem. Rev. 61, 265 (1961).
9. *Thornton, E. K., Thornton, E. R.*: In: Isotope effects in chemical reactions, p. 215 (ed. by *Collins, C. J., Bowman, N. S.*). New York: Van Nostrand Reinhold Co. 1970.
10. *Zollinger, H.*: In: Adv. Phys. Org. Chem., Vol. 2, p. 163 (ed. by *Gold, V.*). London: Academic Press 1964.
11. *Melander, L.*: The use of nuclides in the determination of organic reaction mechanisms, p. 44. Notre Dame: University of Notre Dame Press 1955.
12. *Swain, C. G., Stivers, E. C., Reuwer, Jr., J. F., Schaad, L. J.*: J. Am. Chem. Soc. 80, 5885 (1958).
13. *Olsson, S.*: Arkiv Kemi 16, 489 (1960).
14. *Melander, L., Olsson, S.*: Acta Chem. Scand. 10, 879 (1956).
15. *Melander, L.*: Arkiv Kemi 18, 195 (1961).

M. Orchin and D. M. Bollinger

16. *Gold, V., Tye, F. L.:* J. Chem. Soc. *1952*, 2172.
17. *Olah, G. A., Kuhn, S. J., Pavleth, A.:* J. Am. Chem. Soc. *80*, 6535, 6541 (1958).
18. *Gold, V., Satchell, D. P. N.:* J. Chem. Soc. *1955*, 3609, 3619, 3622.
19. *Breslow, R.:* Organic reaction mechanisms, p. 151. New York: W. A. Benjamin, Inc. 1965.
20. *Mackor, E. L., Hofstra, A., van der Waals, J. H.:* Trans. Faraday Soc. *54*, 66, 186 (1958).
21. *Ogimachi, N., Andrews, L. J., Keefer, R. M.:* J. Am. Chem. Soc. *78*, 2210 (1956).
22. *Brown, H. C., Brady, J. D.:* J. Am. Chem. Soc. *74*, 3570 (1952).
23. *Olah, G. A., Tashiro, M., Kobayashi, S.:* J. Am. Chem. Soc. *92*, 6369 (1970).
24. *Dewar, M. J. S.:* J. Am. Chem. Soc. *74*, 3357 (1952).
25. *Dallinga, G., Verrijn, Stuart, A. A., Smit, P. J., Mackor, E. L.:* Z. Electrochem. *61*, 1019 (1957).
26. *Streitwieser, Jr., A.:* Molecular orbital theory for organic chemists, Chap. 11. New York: John Wiley and Sons 1961.
27. *Streitwieser, Jr., A., Mowery, P. C., Jesaitis, R. G., Lewis, A.:* J. Am. Chem. Soc. *92*, 6529 (1970).
28. *Streitwieser, Jr., A., Lewis, A., Schwager, I., Fish, R. W., Labana, S.:* J. Am. Chem. Soc. *92*, 6525 (1970).
29. *Cavalieri, E., Calvin, M.:* J. Chem. Soc. *1972*, 1253 (Perkin I).
30. *Comyns, A. E., Howald, R. A., Willard, J. E.:* J. Am. Chem. Soc. *78*, 3989 (1956).
31. *Olah, G. A.:* Friedel-Crafts chemistry, p. 420. New York: John Wiley and Sons 1973.
32. *Shatenshtein, A. I., Zhdanova, K. I., Basmanova, V. M.:* Zh. Obshch. Khim. *31*, 250 (1961).
33. *Greenwood, N. N., Wade, K.:* In: Friedel-Crafts and related reactions, Vol. I, p. 581 (ed. by Olah, G.). New York: Interscience Publishers 1963.
34. *Shatenshtein, A. I., Sannikov, A. P., Alikhanov, P. P., Karpov, L. Ya.:* Zh. Obshch, Khim. *35*, 419 (1965).
35. *Sannikov, A. P., Utyanskaya, E. Z., Alikhanov, P. P., Shatenshtein, A. I.:* Zh. Obshch. Khim. *36*, 2036 (1966).
36. *Hawke, D. L., Steigman, J.:* Anal. Chem. *26*, 1989 (1954).
37. *Garnett, J. L., Long, M. A., Vining, R. F. W., Mole, T.:* J. Am. Chem. Soc. *94*, 5913 (1972).
38. *Garnett, J. L., Long, M. A., Vining, R. F. W., Mole, T.:* J. C. S. Chem. Commun. *1972*, 1172.
39. *Shatenshtein, A. I.:* Tetrahedron *18*, 95 (1962).
40. *Setkina, V. N., Barnetskaya, N. K., Anisimov, K. N., Kursanov, D. N.:* Izv. Akad. Nauk SSSR, Ser. Khim. *1964*, 1873; C. A. 62:2684[e].
41. *Setkima, V. N., Kursanov, D. N.:* Russ. Chem. Rev. *37, 1968*, 737.
42. *Nicholls, B., Whiting, M. C.:* J. Chem. Soc. *1959*, 551.
43. *Brown, D. A., Raju, J. R.:* J. Chem. Soc. (A) *1966*, 40.
44. *Nesmeyanov, A. N., Kursanov, D. N., Setkina, V. N., Kislyakova, N. V., Kolobova, N. E., Anisimov, K. N.:* Izv. Akad. Nauk SSSR, Ser. Khim. *1966*, 944.
45. *Trahanovsky, W. S., Card, R. J.:* J. Am. Chem. Soc. *94*, 2897 (1972).
46. *Katz, T. J., Rosenberger, M.:* J. Am. Chem. Soc. *85*, 2030 (1963).
47. *Chatt, J., Davidson, J. M.:* J. Chem. Soc. *1965*, 843.
48. *Parshall, G. W., Knoth, W. H., Schunn, R. A.:* J. Am. Chem. Soc. *91*, 4990 (1969).
49. *Bennett, M. A., Milner, D. L.:* J. Am. Chem. Soc. *91*, 6983 (1969).
50. *Keim, W.:* J. Organometal. Chem. *14*, 179 (1968).
51. *Itatani, H., Bailar, Jr., J. C.:* J. Am. Chem. Soc. *89*, 1600 (1967).
52. *Parshall, G. W.:* Accounts Chem. Res. *3*, 139 (1970). This review is updated in: Collected accounts of transitional metal chemistry. Washington, D. C.: American Chemical Society 1973.
53. *Schunn, R. A.:* Inorg. Chem. *9*, 2567 (1970).
54. *Hodges, R. J., Garnett, J. L.:* J. Phys. Chem. *72*, 1673 (1968).
55. *Hodges, R. J., Garnett, J. L.:* J. Catalysis *13*, 83 (1969).
56. *Hodges, R. J., Garnett, J. L.:* J. Phys. Chem. *73*, 1525 (1969).
57. *Garnett, J. L.:* Catalysis Rev. *5*, 229 (1971).

58. *Davis, K., Garnett, J. L., Hoa, K., Kenyon, R. S., Long, M. A.:* Proc. Vth Internat. Cong. Catalysis, West Palm Beach, Florida (1972) Paper *37*.
59. *Garnett, J. L., West, J. C.:* Aus. J. Chem. *27*, 129 (1974).
60. *Garnett, J. L., Long, M. A., McLaren, A. B., Peterson, K. B.:* J. C. S., Chem. Commun. 1973, 749.
61. *Orchin, M.:* In: Advances in catalysis, Vol. 5 (ed. by *Frankenberg, W. G., Komarewsky, V. I., Rideal, E. K.*). New York: Academic Press 1953.
62. *Friedman, S., Metlin, S., Svedi, A., Wender, I.:* J. Org. Chem. *24*, 1287 (1959).
63. *Taylor, P. D., Orchin, M.:* J. Org. Chem. *37*, 3913 (1972).
64. *Weil, T. A., Friedman, S., Wender, I.:* J. Org. Chem. *39*, 48 (1974).
65. *Wender, I., Levine, R., Orchin, M.:* J. Am. Chem. Soc. *72*, 4375 (1950).

Received October 7, 1974

Structure and Bonding: Index Volume 1-23

A. GOSSAUER
Die Chemie der Pyrrole

17 Abbildungen. XX, 433 Seiten. 1974
(Organische Chemie in Einzeldar-
stellungen, Band 15)
Gebunden DM 158,—; US $68.00
ISBN 3-540-06603-9

Das Pyrrol und seine Derivate haben als
technische Grundstoffe wie auch als
Naturprodukte wachsendes Interesse
gewonnen.

Diese Monographie ist eine umfassende
Übersicht über die seit 1934 erschienene
Literatur (ausgenommen Porphyrine).
Bedingt durch die seitdem ständig
wachsende Anzahl der Veröffentlichungen,
die sich mit den physikalischen Eigenschaften dieser Verbindungsklasse befas-
sen, weichen Konzeption und Gliederung
dieses Buches von denjenigen des klas-
sischen Werkes von H. Fischer und
H. Orth grundsätzlich ab. Die Anwendung
quantenmechanischer Rechenverfahren
zur Deutung der Eigenschaften des
Pyrrol-Moleküls wird im ersten Kapitel
ausführlich erörtert. Die entscheidende
Bedeutung der physikalischen Methoden zur Untersuchung der Konstitution
und Reaktivität des Pyrrols und seiner
Derivate ist durch zahlreiche tabellarisch
geordnete Datenangaben, deren Interpretation im Text diskutiert wird, hervor-
gehoben. Dem präparativ arbeitenden
Chemiker soll die Systematisierung der
synthetischen Methoden bei der Suche
nach der einschlägigen Literatur helfen:
Ringsynthesen sind nach dem Aufbau-
modus des Heterocyclus, die Einführung
von Substituenten nach funktionellen
Gruppen klassifiziert und anhand von
Schemata übersichtlich zusammengefaßt
worden. Bei der Zusammenstellung der
Abbildungen wurden neben den trivialen
Beispielen, die zum besseren Verständnis
des Textes dienen, besonders jene Reaktionen ausgewählt, bei denen Pyrrole
Ausgangsverbindungen zur Darstellung
anderer Heterocyclen (Indole, Pyrrolizine,
Azepine, u.a.) sind. Besondere Sorgfalt
gilt der Beschreibung von Reaktions-
mechanismen. (2621 Literaturzitate.)

M. SCHLOSSER
Struktur und Reaktivität polarer Organometalle

Eine Einführung in die Chemie organischer
Alkali- und Erdalkalimetall-Verbindungen

29 Abbildungen. XI, 187 Seiten. 1973
(Organische Chemie in Einzeldarstellungen,
Band 14)
Gebunden DM 78,—; US $33.60
ISBN 3-540-05719-6

„Mit Organometallen ist nichts unmög-
lich — aber auch nichts voraussagbar",
so lautet, auf einen knappen Nenner
gebracht, ein verbreitetes Vorurteil.
Damit will der Autor aufräumen. Er zeigt,
daß alles mit rechten Dingen zugeht.
Da wird zunächst die Struktur organometallischer Verbindungen — im Kristall
und in Lösung — unter die Lupe genom-
men und als Folge des unbefriedigten
Koordinationsstrebens des Metalls
begreiflich gemacht. Die mangelnde
„Sympathie" zwischen den Bindungs-
partnern Metall und Kohlenstoff vermag
auch die vielen eigenartigen Solvations-
phänomene und das Streben zur Ionen-
paar-Bildung zu erklären. Danach werden
ausführlich die sterischen, induktiven und
mesomeren Effekte behandelt, die über
die Basizität eines Organometalls und
somit dessen „chemisches Potential" ent-
scheiden. Darauf baut dann die ab-
schließende, umfassende Diskussion des
reaktiven Verhaltens organometallischer
Verbindungen auf. Besonders eindrucks-
voll ist das Schlußkapitel, worin gezeigt
wird, wie detaillierte mechanistische
Kenntnisse das Instrumentarium organometallischer Reaktionen vollendet zu
beherrschen erlauben.

Preisänderungen vorbehalten

Springer-Verlag
Berlin Heidelberg New York

CH. K. JØRGENSEN
Oxidation Numbers and Oxidation States
VII, 291 pages. 1969 (Molekülverbindungen und Koordinationsverbindungen in Einzeldarstellungen)
Cloth DM 68,–; US $29.30   ISBN 3-540-04658-5

Contents: Formal Oxidation Numbers. Configurations in Atomic Spectroscopy. Characteristics of Transition Group Ions. Internal Transitions in Partly Filled Shells. Inter-Shell Transitions. Electron Transfer Spectra and Collectively Oxidized Ligands. Oxidation States in Metals and Black Semi-Conductors. Closed-Shell Systems, Hydrides and Back-Bonding. Homopolar Bonds and Catenation. Quanticule Oxidation States. Taxological Quantum Chemistry.

Electrons in Fluids
The Nature of Metal-Ammonia Solutions
Editors: J. Jortner, N.R. Kestner
271 figures. 59 tab. XII, 493 pages. 1973
Cloth DM 120,–; US $51.60   ISBN 3-540-06310-2

Contents: Theory of Electrons in Polar Fluids. Metal-Ammonia Solutions: The Dilute Region. Metal Solutions in Amines and Ethers. Ultrafast Optical Processes. Metal-Ammonia Solutions: Transition Range. The Electronic Structures of Disordered Materials. Concentrated $M-NH_3$ Solutions: A Review. Strange Magnetic Behavior and Phase Relations of Metal-Ammonica Compounds. Metallic Vapors. Mobility Studies of Excess Electrons in Nonpolar Hydrocarbons. Optical Absorption Spectrum of the Solvated Electron in Ethers and Binary Liquid Systems. Subject Index. Color Plates.

Preisänderungen vorbehalten
Prices are subject to change without notice

W. SCHNEIDER
Einführung in die Koordinationschemie
38 Abbildungen. VIII, 173 Seiten. 1968
Gebunden DM 36,–; US $15.50   ISBN 3-540-04324-1

Inhaltsübersicht: Einleitung: Koordination – Ein Gesichtspunkt. Metallatome und Metallionen. Liganda- tome und Liganden. Metallaquoionen. Ligandersatz in wäßriger Lösung: Gleichgewichte. Regelmäßigkeiten und chrakteristische Züge in der Koordinationstendenz metallischer Zentren. Robuste Komplexe und präparative Komplexchemie. Typus und Verlauf von Reaktionen. Anorganische Chromophore. Ligandatome mit niedriger Elektronegativität. Angehäufte Metallatome als Koordi- nationszentren. Bindungstheoretische Ergänzungen. Sach- verzeichnis. Kurzzeichen von Liganden.

Springer-Verlag
Berlin
Heidelberg
New York